ENVIRONMENTAL
FORENSICS
A Glossary of Terms

Robert D. Morrison

CRC Press

Boca Raton   London   New York   Washington, D.C.

Library of Congress Cataloging-in-Publication Data

Morrison, Robert D.
    Environmental forensics : a glossary of terms / by Robert D.
Morrison.
       p. cm.
    ISBN 0-8493-0001-0 (alk. paper)
    1. Environmental law — United States — Dictionaries.
    2. Environmental protection — Dictionaries.   3. Pollutants — Dictionaries.
    I. Title.
    KF3775.A68M67   1999
    344.73'046'03—dc21
                                           99-39051
                                              CIP

No claim to original U.S. Government works
International Standard Book Number 0-8493-0001-0
Library of Congress Card Number 99-39051
Printed in the United States of America 1 2 3 4 5 6 7 8 9 0
Printed on acid-free paper

Preface

This glossary is intended to provide practicing attorneys or consultants with a useful reference to define words and acronyms used in environmental reports, regulatory correspondence, interrogatories, and testimony. Words and acronyms included in this glossary are not discipline specific but rather encompass geologic, toxicological, chemical, and engineering terms whose common thread is their use in environmental forensics. The 3500 definitions and 1500 acronyms were extracted from millions of pages of environmental reports. When a term was encountered that was unfamiliar or new, it was defined and entered into this expanding glossary. It is my hope that this glossary assists you in expediting your review of environmental forensic documents and in gaining a clear understanding of the text.

Best wishes for a successful and informed environmental career.

Robert. D. Morrison
San Diego, CA

The Author

 **Robert Daniel Morrison** has a B.S. in Geology, an M.S. in Environmental Studies, an M.S. in Environmental Engineering, and a Ph.D. in Soil Physics from the University of Wisconsin at Madison. Dr. Morrison has been working for 27 years in the environmental field on issues related to soil and groundwater contamination. He specializes in the forensic review and interpretation of scientific data used in support of litigation involving soil and groundwater contamination. Dr. Morrison has published articles and books on soil and groundwater contamination topics and has shared this information via lectures throughout the world. He is active in reviewing technical papers on forensics techniques and has served on the editorial boards of *Ground Water* and *Groundwater Monitoring Review and Remediation* and currently serves on the editorial board of *The International Journal of Environmental Forensics*. Dr. Morrison has worked as an expert witness and consultant for the U.S. Department of Justice, the Environmental Protection Agency (EPA), and numerous law firms on cases where environmental forensics were used to allocate responsibility. In the capacity as an expert witness and confidential consultant, Dr. Morrison has provided testimony in numerous cases, some with claims ranging from tens of thousands of dollars to as much as five billion dollars.

Acknowledgments

Scientists who directly assisted in the preparation of this book include Sherri Komelyan, Kathleen Calsbeck, Jamie Campos, Kevin Vaughn, and Christian Benitez of R. Morrison & Associates, Inc. Numerous colleagues and researchers provided assistance in the form of communication and information. Special thanks to Dr. Jim Bruya, of Friedman & Bruya in Seattle, WA; Dr. James Szecsody, of Battelle Northwest Laboratories in Hanford, WA; Dr. Blayne Hartman, of TEG in Solano Beach, CA; Kevin Beneteau, of Golder Associates, Calgary, Alberta, Canada; Dr. Barbara Sherwood Lollar, Department of Geology, University of Toronto, Canada; Dawn Zemo, of Geomatrix in San Francisco, CA; and Dr. Ramona Aravena, University of Waterloo, Waterloo, Ontario, Canada.

Special thanks to the wonderful group at CRC Press, especially Becky McEldowney, who provided creative insight and inspiration, and Debrah Goldfarb. who provided marketing direction. Special thanks, too, to Sarah Nicely Fortener of Nicely Creative Services in Geneva, IL, for her wonderful editing of this book.

Table of Contents

Acronyms and Abbreviations

Acronyms and Abbreviations

A

A	=	(1) area; (2) angular (well log)
A/A	=	as above (well log)
AA	=	(1) atomic absorption (spectrometry); (2) administering agency; (3) accountable area/FMSD; (4) adverse action; (5) advice of allowance; (6) assistant administrator; (7) associate administrator
AAAS	=	American Association for the Advancement of Science
AAEE	=	American Academy of Environmental Engineers
AAES	=	American Association of Engineering Societies
AAL	=	applied action level
AAP	=	Asbestos Action Program
AARC	=	Alliance for Acid Rain Control
ABEL	=	EPA computer model to analyze a violator's ability to pay civil penalties
ABES	=	Alliance for Balanced Environmental Solutions
ABIH	=	American Board of Industrial Hygiene
ABMA	=	American Boiler Manufacturers Association
ABS	=	acrylonitrile butadiene styrene
ABTRES	=	abatement and residual forecasting model
Ac	=	actinium
A&C	=	abatement and control
ACA	=	American Conservation Association
ACBM	=	asbestos-containing building material
ACEC	=	American Consulting Engineers Council
ACFM	=	actual cubic feet per minute
ACI	=	American Concrete Institute
ACL	=	alternate concentration limit
ACM	=	asbestos-containing material
ACO	=	Administrative Consent Order
ACP	=	air carcinogen policy
ACQUIRE	=	aquatic information retrieval
ACS	=	American Chemical Society

ACTS	=	Asbestos Contractor Tracking System
ACWA	=	American Clean Water Association
ACWM	=	asbestos-containing waste material
ADABA	=	acceptable database
ADB	=	applications database
ADC	=	Air Diffusion Council
ADI	=	allowable daily intake
ADP	=	automated data processing
ADR	=	alternate dispute resolution
ADSS	=	air data screening system
AEA	=	Atomic Energy Act
AEC	=	Atomic Energy Commission
AED	=	Air Enforcement Division
AEE	=	Alliance for Environmental Education
AEERL	=	Air and Energy Engineering Research Laboratory
AEG	=	Association of Engineering Geologists
AEM	=	acoustic emission monitoring
AEP	=	accepted engineering practices
AERE	=	Association of Environmental and Resource Economists
AES	=	(1) American Electroplating Society; (2) auger electron spectrometry
AESA	=	Association of Environmental Scientists and Administrators
AF	=	acetylaminofluorene
Ag	=	silver
AGA	=	American Gas Association
AGMA	=	American Gear Manufacturer Association
AGU	=	American Geophysical Union
AHERA	=	Federal Asbestos Hazard Emergency Response Act
AHM	=	acutely hazardous materials
A_i	=	area of Phase I (m²)
A&I	=	alternative and innovative wastewater treatment system
AIC	=	(1) acceptable intake for chronic exposure (mg/kg/day); (2) active to inert conversion
AICE	=	American Institute of Chemical Engineers
AIH	=	American Institute of Hydrology
AIHC	=	American Industrial Health Council
AIP	=	auto ignition point
AIPG	=	American Institute of Professional Geologists
AIRS	=	(1) Aerometric Information Retrieval System; (2) air quality subsystem
AIS	=	asbestos information system
AISC	=	American Institute of Steel Construction
AISI	=	American Iron and Steel Institute

Al	=	aluminum
AL	=	acceptable level
ALA	=	delta-aminolevulinic acid
ALAPCO	=	Association of Local Air Pollution Control Officials
ALARS	=	as low as reasonably achievable
ALAS	=	Anthropogenic Lead Archeostatigraphy Model
ALC	=	application-limiting constituent
ALJ	=	administrative law judge
ALMS	=	atomic line molecular spectroscopy
ALPM	=	automated log-P measurement
ALT	=	altered (well log)
ALTA	=	American Land Title Association
Am	=	americium
AMBIENS	=	atmospheric mass balance of industrially emitted and natural sulfur
AMC	=	Army Material Command
AMCA	=	Air Moving and Conditioning Association
AMIS	=	air management information system
AMOS	=	air management oversight system
AMSA	=	Association of Metropolitan Sewer Agencies
AMSL	=	above mean sea level
ANILCA	=	Alaska National Interest Lands Conservation Act
ANPR	=	advance notice of proposed rulemaking
ANRHRD	=	Air, Noise, and Radiation Health Research Division
ANSI	=	American National Standards Institute (formerly U.S.A. National Standards Institute)
AO	=	(1) administrative officer; (2) administrative order; (3) awards and obligations
AOAC	=	Association of Official Analytical Chemists
AOD	=	argon-oxygen decarbonization
AOML	=	Atlantic Oceanographic and Meteorological Laboratory
AP	=	accounting point
APA	=	(1) Administrative Procedures Act; (2) American Planning Association
APCA	=	Air Pollution Control Association
APCD	=	Air Pollution Control District
APDS	=	Automated Procurement Documentation System
APER	=	Air Pollution Emissions Report
APHA	=	American Public Health Association
APHNT	=	aphanitic (well log)
API	=	(1) American Petroleum Institute; (2) American Paper Institute
APP	=	approximately (well log)

APPA	=	American Public Power Association
APRAC	=	urban diffusion model for carbon monoxide from motor vehicle traffic
APTI	=	Air Pollution Training Institute
APWA	=	American Public Works Association
AQ-7	=	non-reactive pollutant modeling
AQCR	=	Air Quality Control Region
AQDM	=	air quality display model
AQMA	=	air quality maintenance area
AQMD	=	air quality management district
AQSM	=	air quality simulation model
AQTAD	=	air quality technical assistance demonstration
AR	=	administrative record
Ar	=	argon
ARA	=	(1) assistance regional administrator: (2) associate regional administrator
ARAR	=	applicable or relevant and appropriate standards, limitations, criteria, and requirements
ARB	=	Air Resources Board
ARC	=	Agency ranking committee
ARCC	=	American Rivers Conservation Council
ARCS	=	alternative remedial contract strategy
AREA	=	American Railway Engineer Association
ARG	=	American Resources Group
ARIP	=	Accidental Release Information Program
ARL	=	Air Resources Laboratory
ARM	=	Air Resources Management
ARMOS	=	a real multiphase organic simulator
ARO	=	alternate regulatory option
ARPO	=	Acid Rain Policy Office
ARRP	=	acid rain research program
ARRPA	=	Air Resources Regional Pollution Assessment Model
ARS	=	Agricultural Research Service
ARZ	=	Auto Restrict Zone
AS	=	area source
As	=	arsenic
ASC	=	area source category
ASCE	=	American Society of Civil Engineers
ASCII	=	American Standard Code for Information Interchange
ASCL	=	acceptable soil contaminant levels
ASCP	=	American Society of Consulting Planners
ASDWA	=	Association of State Drinking Water Administrators
ASHAA	=	Asbestos in Schools Hazard Abatement Act of 1984

ASHRAE	=	American Society of Heating, Refrigerating and Air Conditioning Engineers
ASIWPCA	=	Association of State and Interstate Water Pollution Control Administrators
ASMDHS	=	airshed model data handling system
ASME	=	American Society of Mechanical Engineers
ASPIS	=	Abandoned Site Program Information System Facility Profile Reports (State of California Department of Health Services)
ASRL	=	Atmospheric Sciences Research Laboratory
ASSE	=	American Society of Sanitary Engineers
AST	=	aboveground storage tank
ASTHO	=	Association of State and Territorial Health Officials
ASTM	=	American Society for Testing and Materials
ASTSWMO	=	Association of State and Territorial Solid Waste Management Officials
AT	=	(1) advanced treatment; (2) alpha track detection
At	=	astatine
ATERIS	=	Air Toxics Exposure and Risk Information System
ATMI	=	American Textile Manufacturing Institute
ATS	=	action tracking system
ATSDR	=	Agency for Toxic Substances and Disease Registry
ATTF	=	Air Toxics Task Force
AU	=	auger sample
Au	=	gold
AUSM	=	Advanced Utility Simulation Model
AVS/SEM	=	Acid volatile sulfide/sequential extraction metals
AWMA	=	Air and Waste Management Association
AWPA	=	American Wood-Preservers' Association
AWQC	=	(1) Aquatic Water Quality Criteria; (2) Federal Water Quality Criteria
AWRA	=	American Water Resources Association
AWS	=	American Welding Society
AWT	=	Advanced Wastewater Treatment
AWWA	=	American Water Works Association
AWWARF	=	American Water Works Association Research Foundation

B

b	=	saturated thickness of aquifer or formation
B	=	boron
Ba	=	barium

BAA	=	Board of Assistance Appeals
BACM	=	best available control measures
BACT	=	best available control technology
BADT	=	best available demonstrated technology
BAF	=	bioaccumulation factor
BaP	=	benzo(a)pyrene
BARF	=	best available retrofit facility
BART	=	best available retrofit technology
BAT	=	(1) best available technology; (2) best available treatment
BATEA	=	best available treatment economically achievable
BCCM	=	Board for Certified Consulting Meteorologists
BCF	=	bioconcentration factor
BCPCT	=	best conventional pollutant control technology
BCT	=	best control technology
BDAT	=	best demonstrated available technology
BDCT	=	best demonstrated control technology
BDT	=	best demonstrated technology
Be	=	beryllium
BED	=	Board of Expenditure (State of California Superfund List)
BEJ	=	best engineering judgment
BF	=	*bona fide* notice of intent to manufacture or import
BG	=	billion gallons
BGS	=	below ground surface
BHC	=	benzene hexachloride
Bhp	=	brake horsepower
BI	=	background information
Bi	=	bismuth
BIA	=	Bureau of Indian Affairs
BID	=	(1) background information document; (2) buoyancy-induced dispersion
BIF	=	boilers and industrial furnaces
BIOPLUME	=	model to predict the maximum extent of existing plumes
Bk	=	berkelium
BLDR	=	boulder/bouldery (well log)
BLEVE	=	boiling liquid expanding vapor explosions
BLK	=	black (well log)
BLM	=	Bureau of Land Management
BLS	=	Bureau of Labor Statistics
BMDL	=	below method detection level
BMP	=	best management practice
BMSL	=	below mean sea level
BNA	=	(1) Bureau of National Affairs; (2) base neutral analysis
BOA	=	basic ordering agreements

BOD	=	(1) biological oxygen demand; (2) biochemical oxygen demand
BOF	=	basic oxygen furnace
BOH	=	bottom of hole
BOM	=	Bureau of Mines
BOP	=	basic oxygen process
BOPF	=	basic oxygen process furnace
BPHA	=	baseline public health assessment
BPJ	=	best professional judgment
BPT	=	best practicable treatment
BPWTT	=	best practical wastewater treatment technology
Br	=	bromine
BRS	=	bibliographic retrieval service
BS	=	bag sample
BSI	=	British Standards Institute
BSLT	=	basalt (well log)
BSO	=	benzene-soluble organics
Btu	=	British thermal unit
BTX	=	benzene, toluene, and xylene
BTXE	=	benzene, toluene, xylene, and ethylbenzene
BUN	=	blood urea nitrogen
BW	=	body weight

C

C	=	(1) carbon; (2) concentration coarse (well log); (3) degrees centigrade; (4) degrees Celsius; (5) molar concentration (mol/L or mmol/m^3)
^{14}C	=	radioactive-labeled carbon-14 compound
$CClF_3$	=	monochlorotrifluoromethane (Freon-13)
CCl_2F_2	=	dichlorodifluoromethane (Freon-12)
CCl_3F	=	trichloromonofluoromethane (Freon-11) (Carrene-2)
CCl_4	=	carbon tetrachloride
$C_2Cl_2F_4$	=	dichlorotetrafluoroethane (Freon-114)
$C_2Cl_3F_3$	=	trichlorotrifluoroethane (Freon-113)
C_2Cl_4	=	tetrachloroethylene
C_2Cl_6	=	hexachloroethane
C_4C_{12}	=	C_4 = 4 carbons in a chain (e.g., $CH_3CH_2CH_2CH_3$); C_{12} = 12 carbons in a chain
CF_3Br	=	bromotrifluoromethane
CF_4	=	tetrafluoromethane
$CHClF_2$	=	monochlorodifluoromethane (Freon-22)

$CHCl_2F$	=	dichloromonofluoromethane (Freon-21)
$CHCl_3$	=	chloroform
CH_2O	=	formaldehyde
CH_3CHF_2	=	ethylidene fluoride 26.2% (Carrene-7)
CH_3Cl	=	methyl chloride
CH_3I	=	methyl iodide
$(CH_3)_3CH$	=	isobutane
C_2HCl_3	=	trichloroethylene
$C_2H_2Cl_2$	=	1,1-dichloroethane
$C_2H_2Cl_4$	=	1,1,2,2-tetrachloroethane
C_2H_3Cl	=	vinyl chloride
$C_2H_3Cl_3$	=	1,1,1-trichloroethane
C_2H_3ClO	=	acetyl chloride
C_2H_4	=	ethylene
$C_2H_4Br_2$	=	1,2-dibromoethane
$C_2H_4Cl_2$	=	1,2-dichloroethane
C_2H_4ClO	=	*bis*(chloromethyl)ether
C_2H_4O	=	acetaldehyde
C_2H_4O	=	ethylene oxide
C_2H_6	=	ethane
C_3H_3N	=	acrylonitrile
C_3H_4O	=	acrolein
$C_3H_4O_2$	=	acrylic acid
C_3H_5ClO	=	epichlorohydrin
C_3H_5NO	=	acrylamide
C_3H_6	=	propylene
C_3H_6O	=	1,2-propylene oxide
C_3H_6O	=	allyl alcohol
C_3H_7NO	=	N,N-dimethylformamide
C_3H_8	=	propane
C_3H_9N	=	trimethylamine
$C_4H_2O_3$	=	maleic anhydride
$C_4H_4O_4$	=	maleic acid
C_4H_6	=	1,3-butadiene
$C_4H_6O_2$	=	vinyl acetate
$C_4H_8Cl_2O$	=	*bis*(2-chloroethyl)ether
C_4H_8O	=	tetrahydrofuran
$C_4H_8O_2$	=	1,4-dioxane
C_4H_{10}	=	butane
$C_5H_8O_2$	=	methylmethacrylate
C_5H_{12}	=	*n*-pentane
C_6Cl_6	=	hexachlorobenzene
$C_6H_3Cl_3$	=	1,2,4-trichlorobenzene

$C_6H_3Cl_3$	=	1,3,5-trichlorobenzene
$C_6H_3Cl_3O$	=	2,4,6-trichlorophenol
$C_6H_4Cl_2$	=	*m*-dichlorobenzene
$C_6H_4Cl_2$	=	*o*-dichlorobenzene
$C_6H_4Cl_2$	=	*p*-dichlorobenzene
$C_6H_4ClNO_2$	=	*o*-chloronitrobenzene
C_6H_5Cl	=	chlorobenezene
C_6H_5ClO	=	*m*-chlorophenol
C_6H_5ClO	=	*o*-chlorophenol
C_6H_5ClO	=	*p*-chlorophenol
$C_6H_5NO_2$	=	nitrobenzene
$C_6H_5NO_3$	=	*m*-nitrophenol
$C_6H_5NO_3$	=	*o*-nitrophenol
$C_6H_5NO_3$	=	*p*-nitrophenol
C_6H_6	=	benzene
C_6H_6ClN	=	*o*-chloroaniline
C_6H_6ClN	=	*p*-chloroaniline
C_6H_6O	=	phenol
C_6H_7N	=	aniline
$C_6H_{10}O_4$	=	adipic acid
$C_6H_{11}NO$	=	caprolactam
C_6H_{12}	=	cyclohexane
C_6H_{14}	=	*n*-hexane
$C_7H_5Cl_3$	=	trichloromethylbenzene
$C_7H_6O_2$	=	benzoic acid
C_7H_7Cl	=	chloromethylbenzene
$C_7H_7NO_2$	=	*o*-nitrotoluene
$C_7H_7NO_2$	=	*p*-nitrotoluene
C_7H_8	=	toluene
C_7H_8O	=	*m*-cresol
C_7H_8O	=	*o*-cresol
C_7H_8O	=	*p*-cresol
$C_7H_{10}N_2$	=	toluene-2,4-diamine
C_7H_{16}	=	*n*-heptane
$C_8H_4O_3$	=	phthalic anhydride
C_8H_8	=	styrene
C_8H_{10}	=	ethylbenzene
C_8H_{10}	=	*m*-xylene
C_8H_{10}	=	*o*-xylene
C_8H_{10}	=	*p*-xylene
$C_8H_{10}O$	=	2,4-xylenol
$C_9H_6N_2O_2$	=	toluene-2,6-diisocyanate
$C_9H_6N_2O_2$	=	toluene-2,4-diisocyanate

$C_{10}H_8$	=	naphthalene
$C_{10}H_{10}O_4$	=	dimethyl phthalate
$C_{12}H_{12}N_2$	=	benzidine
$C_{12}H_{14}O_4$	=	diethyl phthalate
$C_{16}H_{22}O_4$	=	dibutyl phthalate
CO	=	carbon monoxide
CO_2	=	carbon dioxide
CS_2	=	carbon disulfide
CA	=	(1) citizen act; (2) competition advocate; (3) cooperative agreements; (4) corrective action; (5) cost analysis
Ca	=	calcium
C_A	=	concentration in air phase (mol/L or mmol/m³)
CAA	=	(1) Federal Clean Air Act; (2) Compliance Assurance Agreement
CAAA	=	Federal Clean Air Act Amendments
CAB	=	Civil Aeronautics Board
CABF	=	Cochran's approximation to Behrens-Fisher t-test
CAC	=	California Administrative Code
CAER	=	Community Awareness and Emergency Response
CAFO	=	Consent Agreement/Final Order
CAG	=	Carcinogenic Assessment Group (EPA)
CAGE	=	Commercial and Government Entity
CAGI	=	Compressed Air and Gas Institute
CAH	=	chlorinated aliphatic hydrocarbons
CAIR	=	Comprehensive Assessment of Information Rule
Cal/OSHA	=	State of California Department of Industrial Relations, Division of Industrial Safety
CALINE	=	California Line Source Model
CAMP	=	Continuous Air Monitoring Program
CAO	=	Corrective Action Order
CAP	=	(1) correction action plan; (2) cost allocation procedure; (3) criteria air pollutant
CAPA	=	critical aquifer protection area
CAR	=	corrective action report
CAS No.	=	Chemical Abstract Services Number
CASAC	=	Clean Air Scientific Advisory Committee
CASETRK	=	FIFRA and TSCA Case Tracking System
CATS	=	Corrective Action Tracking System
CAU	=	carbon adsorption unit
CB	=	(1) chlorobenzene; (2) carbide bit; (3) continuous bubbler
CBBL	=	cobble/cobbly (well log)
CBD	=	Commerce Business Daily
CBEP	=	community-based environmental project

CBI	=	(1) compliance biomonitoring inspection; (2) confidential business information
CBM	=	certified ballast manufacturer
CBO	=	Congressional Budget Office
CBOD	=	carbonaceous biochemical oxygen demand
CBR	=	California Bearing Ratio
CC	=	activated charcoal adsorption
CC/RTS	=	Chemical Collection/Request Tracking System
CCAA	=	Canadian Clean Air Act
CCEA	=	Conventional Combustion Environmental Assessment
CCID	=	Confidential Chemical Identification System
CCR	=	California Code of Regulations
CCRIS	=	Chemical Carcinogenesis Research Information System
CCTP	=	Clean Coal Technology Program
Cd	=	cadmium
CDD	=	chlorinated dibenzo-p-dioxin
CDF	=	chlorinated dibenzofuran
CDI	=	chronic daily intake exposure
CDS	=	Compliance Data System
CDWR	=	California Department of Water Resources
CE	=	(1) categorical exclusion; (2) conditionally exempt generator
Ce	=	cerium
CEARC	=	Canadian Environmental Assessment Research Council
CEB	=	chemical element balance
CECATS	=	CSB Existing Chemicals Assessment Tracking System
CEEM	=	Center for Energy and Environmental Management
CEG	=	Certified Engineering Geologist
CEI	=	compliance evaluation inspection
CEM	=	continuous emissions monitoring
CEPA	=	Canadian Environmental Protection Act
CEPP	=	Federal Chemical Emergency Preparedness Plan
CEQ	=	Council on Environmental Quality
CERCLA	=	Comprehensive Environmental Response, Compensation, and Liability Act of 1980 (Superfund)
CERCLIS	=	Comprehensive Environmental Response, Compensation, and Liability Information System
Cf	=	californium
CFC	=	chlorofluorocarbons
CFEST	=	coupled fluid, energy, and solute transport
CFM	=	chlorofluoromethanes
cfm	=	cubic feet per minute (also ft³/min)
CFR	=	Code of Federal Regulations
cfs	=	cubic feet per second

CGI	=	combustible gas indicator
CGL	=	comprehensive general liability
cgl	=	conglomerate
CGMEs	=	comprehensive groundwater monitoring evaluations
ch	=	chert
CH	=	inorganic clays of high plasticity
CHAMP	=	Community Health Air Monitoring Program
CHC	=	chlorinated hydrocarbons
CHEMNET	=	Chemical Industry Emergency Mutual Aid Network
CHESS	=	Community Health and Environmental Surveillance System
CHIP	=	Chemical Hazard Information Profile (OPTS)
CHIPS	=	Chemical Hazard Information Profile System
CI	=	confidence interval
CIAQ	=	Council on Indoor Air Quality
CIBL	=	Convective Internal Boundary Layer
CICIS	=	Chemicals in Commerce Information System
CIDRS	=	Cascade Impactor Data Reduction System
CIIT	=	Chemical Industry Institute of Toxicology
CIS	=	Chemical Information System
cis-DCE	=	cis-dichloroethylene
CISPI	=	Cast Iron Soil Pipe Institute
CKD	=	cement kiln dust
CKRC	=	Cement Kiln Recycling Coalition
Cl	=	chlorine
C_L	=	subcooled liquid concentration (mol/L or mmol/m^3)
CLC	=	capacity-limiting constituents
CLEANS	=	Clinical Laboratory for Evaluation and Assessment of Toxic Substances
CLEVER	=	Clinical Laboratory for Evaluation and Validation of Epidemiologic Research
CLFM	=	chloroform
CLIPS	=	Chemical List Index and Processing System
CLP	=	Contract Laboratory Program (Superfund)
CLST	=	cluster (well log)
CLY	=	clay/clayey (well log)
cm	=	centimeter
CM	=	corrective measure
Cm	=	curium
cm^3	=	cubic centimeters
CMA	=	Chemical Manufacturers Association
CMB	=	chemical mass balance
CME	=	Comprehensive Monitoring Evaluation
CMEL	=	Comprehensive Monitoring Evaluation Log

CMMA	=	Crane Manufacturers Association of America
CMOS	=	complementary metal oxide semiconductor
CMP	=	Chemical Mechanical Polish
CMPT	=	component (well log)
CMS	=	corrective measures study
CMTD	=	cemented (well log)
CNSLDT	=	consolidated (well log)
Co	=	cobalt
COCO	=	contractor-owned/contractor-operated
COD	=	chemical oxygen demand
COE	=	U.S. Army Corps of Engineers
COH	=	coefficient of haze
COLIWASA	=	composite liquid waste sampler
Consol	=	consolidation
COPC	=	chemical of potential concern
CORPS	=	U.S. Army Corps of Engineers
CP	=	cone penetrometer
CPAF	=	(1) cost plus award fee; (2) carcinogenic polyaromatic hydrocarbon
CPE	=	chlorinated polyethylene
CPF	=	carcinogenic potency factor
CPFF	=	cost plus fixed fee
CPG	=	Certified Professional Geologist
CPIF	=	cost plus incentive fee
cpm	=	counts per minute
CPO	=	Certified Project Officer
CPU	=	central processing unit
Cr	=	chromium
CR	=	continuous radon monitoring
CRDL	=	contract-required detection limit
CREEL	=	Cold Regions Research Engineering Laboratory
CRP	=	(1) Consolidated Rules of Practice (also CROP); (2) community relations plan
CRSI	=	Concrete Reinforcing Steel Institute
CRSTER	=	single source dispersion model
Cs	=	cesium (also caesium)
C^s	=	saturated aqueous concentration (mol/L or mmol/m^3)
C_s	=	solid molar concentration (mol/L or mmol/m^3)
CSAMT	=	controlled source audio magneto tellurics (surface geophysical technique)
CSCT	=	Consortium for Site Characterization Technology
CSG	=	Council of State Governments
CSI	=	compliance sampling inspection

CSIN	=	Chemical Substances Information Network
CSMA	=	Chemical Specialties Manufacturers Association
CSO	=	combined sewer overflow
CSPE	=	chlorosulfonated polyethylene
CTI	=	Cooling Tower Institute
Cu	=	copper
cu. yds.	=	cubic yards
C_W	=	concentration in water phase (mol/L or mmol/m^3)
CWA	=	Federal Clean Water Act
CWG	=	Community Work Group
CWTC	=	Chemical Waste Transportation Council
CZMA	=	Coastal Zone Management Act

D

d	=	day
D	=	D values (mol/Pa·hr)
D_A	=	D values for advection (mol/Pa·hr)
D_{Ai}	=	D value for advective loss in phase i (mol/Pa·hr)
D_{ij}	=	intermedia D value for air/water diffusion (adsorption) (mol/Pa·hr)
D_{QS}	=	D value for wet and dry deposition (air/soil) (mol/Pa·hr)
D_R	=	D value for reaction loss in phase i (mol/Pa·hr)
D_{Ri}	=	intermedia D values (mol/Pa·hr)
D_{RS}	=	D value for rain dissolution (air/soil) (mol/Pa·hr)
D_{RW}	=	D value for total particle transport (dry and wet) (mol/Pa·hr)
D_S	=	D value for air/soil boundary layer diffusion (mol/Pa·hr)
D_{SA}	=	D value for air transport in soil (mol/Pa·hr)
D_{SW}	=	D value for water transport in soil (mol/Pa·hr)
D_{Ti}	=	total transport D value in bulk phase i (mol/Pa·hr)
D_{VS}	=	D value for total soil/air transport (mol/Pa·hr)
D_{VW}	=	intermedia D value for air/water dissolution (mol/Pa·hr)
DAF	=	dilution attenuation factor
DAR	=	(1) Defense Acquisition Regulations; (2) damage assessment report
dB	=	decibel
DB	=	diamond bit
DCA	=	1,2-dichloroethane
DCB	=	1,4-dichlorobenzene
DCE	=	1,1-dichloroethylene
DCM	=	dichloromethane (methylene chloride)
DCP	=	direct current plasma

Dd	=	dry density
DDT	=	dichloro diphenyl trichloroethane
DE	=	daily exposure
D.E.E.	=	Diplomate of Environmental Engineering
DEHP	=	*bis* (2-ethyl-hexyl) phthalate
DEIS	=	Draft Environmental Impact Statement
DEL	=	dose equivalent limits
DEP	=	Department of Environmental Protection
DEQ	=	Department of Environmental Quality
DEQE	=	Department of Environmental Quality and Engineering
DER	=	Department of Environmental Resources
DERs	=	data evaluation records
DES	=	diethylstilbesterol
DFM	=	diesel fuel marine
DFPA	=	Douglas Fir Plywood Association
DG	=	decomposed granite (well log)
DHS	=	California Department of Health Services
DIMM	=	dual in-line memory module
DLM	=	designated level methodology
DLTNCY	=	dilatancy (well log)
DMDEL	=	dimethyldiethyl lead
DMI	=	Dunn's Market Identifiers
DMP	=	damp (well log)
DMR	=	discharge monitoring report
DNA	=	deoxyribonucleic acid
DNAPL	=	dense non-aqueous phase liquid
DNR	=	Department of Natural Resources
DNS	=	dense (well log)
DO	=	dissolved oxygen
DOC	=	dissolved organic carbon
DOD	=	U.S. Department of Defense
DOE	=	U.S. Department of Energy
DOJ	=	U.S. Department of Justice
DOT	=	U.S. Department of Transportation
DQO	=	data quality objective
DRAM	=	Dynamic Random Access Memory
DRE	=	destruction/removal efficiency
DRK	=	dark (well log)
DS	=	dichotomous sampler
DSCD	=	consolidated drained direct shear test
DSCF	=	dry standard cubic feet
DSCM	=	dry standard cubic meter
DTSC	=	California Department of Toxic Substances Control

DWEL	=	drinking water equivalent level
DWPL	=	drinking water priority list (SARA)
DWS	=	drinking water standard
Dy	=	dysprosium

E

E	=	Emission rate (mol/hr or kg/hr)
EA	=	(1) environmental assessment; (2) endangerment assessment; (3) environmental action; (4) environmental analysis; (5) environmental audit
EAF	=	electric arc furnace; environmental attenuation factor
EAG	=	Exposure Assessment Group (EPA)
EAP	=	environmental action plan
EBCDIC	=	extended binary coded decimal interchange code
EC	=	emulsifiable concentrate
ECD	=	electron capture detector
ECL	=	Environmental Chemical Laboratory
ECR	=	Enforcement Case Review
ECRA	=	Economic Cleanup Responsibility Act
ED	=	effective dose
EDA	=	Economic Development Administration
EDB	=	ethylene dibromide
EDC	=	ethylene dichloride
EDD	=	Enforcement Decision Document
EDF	=	Environmental Defense Fund
EDO	=	extended data out
EDRS	=	Enforcement Document Retrieval System
EDTA	=	ethylene diamine triacetic acid
EEA	=	energy and environmental analysis
EEG	=	electroencephalogram
EER	=	Environmental Effects Report
EERL	=	Eastern Environmental Radiation Laboratory
EERU	=	Environmental Emergency Response Unit
EFR	=	external floating roof
EGR	=	exhaust gas recirculation
EH	=	redox potential
EHS	=	extremely hazardous substances
EI	=	emissions inventory
EIA	=	(1) enzyme immunoassay; (2) ethylene interpolymer alloy; (3) Electronic Industries Association; (4) Environmental Impact Assessment

EIL	=	Environmental Impairment Liability
EIR	=	Environmental Impact Report
EIS	=	Environmental Impact Statement
EKMA	=	empirical kinetic modeling approach
EL	=	exposure level
ELCD	=	electrolytic conductivity detector
ELI	=	Environmental Law Institute
ELISA	=	enzyme-linked immunosorbent assay
ELNGD	=	elongated (well log)
ELR	=	Environmental Law Reporter
ELSTC	=	elastic (well log)
EM	=	electron microscope
EMAS	=	Enforcement Management and Accountability System
EMR	=	Environmental Management Report
EMS	=	Enforcement Management System
EMSL	=	Environmental Monitoring and Support Laboratory (EPA)
EMTS	=	environmental monitoring testing site
EO	=	ethylene oxide
EOF	=	Emergency Operations Facility
EOX	=	extractable organohalogens
EP	=	extraction procedure
EPA	=	U.S. Environmental Protection Agency
EPCRA	=	Emergency Planning and Community Right to Know Act (SARA Title III)
EPD	=	Emergency Planning District
EPDM	=	ethylene propylene diene monomer
EPI	=	Environmental Policy Institute
EPICS	=	equilibrium partitioning in closed system
EPO	=	environmental professional organization
ePTFE	=	expanded polytetrafluoroethene
EP-TOX	=	extraction procedure toxicity test
ER	=	electrical resistivity
Er	=	erbium
ERA	=	expedited response action
ERAMS	=	Environmental Radiation Ambient Monitoring System
ERC	=	(1) Emergency Response Commission; (2) Environmental Research Center; (3) Emissions Reduction Credit
ERCS	=	Emergency Response Cleanup Services
ERD	=	Environmental Restoration Division
ERDA	=	Energy Research and Development Administration
ERL	=	Environmental Research Laboratory
ERP	=	Enforcement Response Policy
ERT	=	emergency response team

ERWM	=	Environmental Restoration and Waste Management
Es	=	einsteinium
ESA	=	environmental site audit
ESCA	=	electron spectroscopy for chemical analysis
ESECA	=	Energy Supply and Environmental Coordination Act
ESP	=	Electrostatic precipitator
ETL	=	(1) Electronic Industries Association; (2) electrical testing laboratory
ETP	=	Emerging Technologies Program
Eu	=	europium

F

f	=	fugacity (Pa)
f_i	=	fugacity in pure phase i (Pa)
F	=	(1) fluorine; (2) fine (well log); (3) fugacity ratio
°F	=	degrees Fahrenheit
F/M	=	food-to-microorganism ratio
FAA	=	Federal Aviation Administration
FASB	=	Financial Accounting Standards Board
FBC	=	fluidized bed combustion
FBI	=	fluidized bed incinerator
FCC	=	fluid catalytic converter
FCCU	=	fluid catalytic cracking unit
f-cost.	=	fragmental constants
FDA	=	Food and Drug Administration
FE	=	fugitive emissions
Fe	=	iron
FED	=	field emission display
FEIS	=	Fugitive Emissions Information System
FEL	=	Frank effect level
FELD	=	feldspar (well log)
FEMA	=	Federal Emergency Management Agency
FEPCA	=	Federal Environmental Pesticides Control Act
FERC	=	Federal Energy Regulatory Commission
FEVI	=	front-end volatility index
FFCA	=	Federal Facility Compliance Agreement
FFFSG	=	fossil-fuel-fired steam generator
FFIS	=	Federal Facilities Information System
FGD	=	flue-gas desulfurization
FHC	=	fuel hydrocarbon
FID	=	flame ionization detector

FIFRA	=	Federal Insecticide, Fungicide, and Rodenticide Act
FIM	=	friable insulation material
FINDS	=	Facility Index System
FIP	=	final implementation plan
FIT	=	field investigation team
FLETC	=	Federal Law Enforcement Training Center
FLPMA	=	Federal Land Policy and Management Act
fluo.	=	fluorescence method
FM	=	Factory Mutual Insurance Company
Fm	=	fermium
FML	=	flexible membrane liner
FMP	=	facility management plan
FMS	=	firm (well log)
FNSI	=	finding of no significant impact
FOIA	=	(1) Freedom of Information Act; (2) fiberoptic immunoassay
FOISD	=	fiberoptic isolated spherical dipole antenna
FONSI	=	finding of no significant impact
FP	=	fine particulate (well log)
FPA	=	Federal Pesticide Act
FPD	=	flame photometric detector
FPEIS	=	Fine Particulate Emissions Information System
FPPA	=	Federal Pollution Prevention Act
FPS	=	Fluid Power Society
Fr	=	francium
FRA	=	Federal Register Act
FRCD	=	fractured (well log)
FREDS	=	Flexible Regional Emissions Data System
FRM	=	firm (well log)
FRN	=	Final Rulemaking Notice
FRV	=	facility reference values
FS	=	(1) federal specifications; (2) feasibility study (CERCLA related)
FS/CAOP	=	feasibility study/corrective action option plan
FSR	=	field sample record
FSSD	=	fissured (well log)
ft	=	feet or foot
ft^2	=	square feet
ft^3	=	cubic feet
FTIR	=	Fourier transform infrared spectroscopy
ft-lb	=	foot-pound
FURS	=	Federal Underground Injection Control Reporting System
FWPCA	=	Federal Water Pollution Control Act
FWS	=	U.S. Fish and Wildlife Service

G

g	=	gram; 1/1000 of a kilogram
G	=	Advective inflow (mol/hr)
G_B	=	Advective inflow to bottom sediment (mol/hr)
G_s	=	specific gravity
ΔG_V	=	Gibbs' free energy of vaporization (kJ/mol or kcal/mol)
Ga	=	gallium
GAC	=	granular activated carbon
GACT	=	granular activated carbon treatment
gal	=	gallons
GAO	=	General Accounting Office
GAS	=	gasoline (well log)
GBP	=	gravity-based penalty
GC	=	(1) gas chromatography; (2) clayey gravels (well log)
GC/ECD	=	gas chromatography analysis with electron capture detector
GC/FID	=	gas chromatography/flame ionization detector
GC/MS	=	gas chromatography/mass spectrometry
GC/PID	=	gas chromatography/photoionization detector
GC-RT	=	gas chromatography retention time
GCWR	=	gross combination weight rating
Ge	=	germanium
GEMS	=	(1) Global Environmental Monitoring System; (2) Graphical Exposure Modeling System
gen. col.	=	generator-column
GFF	=	glass fiber filter
GIS	=	general information submission
GIS	=	geographic information systems
μg/L	=	micrograms per liter (10^{-6})
GLC	=	gas liquid chromatography
GLERL	=	Great Lakes Environmental Research Laboratory
GM	=	silty gravels (well log)
g/mi	=	grams per mile
GMT	=	Greenwich Mean Time
GO 95	=	General Order No. 95, California Public Utilities Commission, Rules for Overhead Electric Line Construction
GO 128	=	General Order No. 128, California Public Utilities Commission, Rules for Underground Electric Line Construction
GOCO	=	government-observed, contractor-operated
GP	=	(1) poorly sorted soil; (2) poorly graded gravel (well log)
GPAD	=	gallons per acre per day
gpd	=	gallons per day
gpd/ft	=	gallons per day per foot

gpd/ft^2	=	gallons per day per square foot
gpg	=	grams per gallon
GPLC	=	gel permeation liquid chromatography
gpm	=	gallons per minute
GPR	=	ground-penetrating radar
GPS	=	groundwater protection strategy
GR	=	(1) grain (well log); (2) grab radon sampling
GRAMs	=	Groundwater Risk Assessment Methodologies
GRDD	=	graded (well log)
GRN	=	green (well log)
GRNTC	=	grantic (well log)
GRVL	=	gravel (well log)
GRY	=	gray (well log)
GS	=	Geological Survey (usually state, sometimes used synonymously with USGS)
GSA	=	(1) Geological Society of America; (2) General Services Administration
GVP	=	gasoline vapor pressure
GW	=	(1) well-graded soil; (2) well-graded (well log); (3) groundwater
GWM	=	(1) groundwater monitoring; (2) Groundwater Policy and Management Staff
GWPS	=	groundwater protection standard
GWQAP	=	groundwater quality assessment plan
GWTF	=	groundwater task force

H

H	=	(1) hydrogen; (2) high (well log)
H, HLC	=	Henry's Law constant (Pa m^3/mol)
ΔH_{fus}	=	enthalpy of fusion (kcal/mol)
ΔH_V	=	enthalpy of vaporization (kJ/mol or kcal/mol)
H$_z$	=	hertz
HAAB	=	Hazard Abatement and Assistance Branch
HAF	=	halogen acid furnaces
HAP	=	hazardous air pollutant
HAPEMS	=	Hazardous Air Pollutant Enforcement Management System
HAR	=	hydrogeological assessment report
HAs	=	health advisories
HAS	=	hollow stem auger
HASP	=	Health and Safety Plan; horizontal air sparging
HASTREMS	=	Hazardous and Trace Emissions System

HAZMAT	=	hazardous materials
HAZWOPER	=	Hazardous Waste Operations and Emergency Response Regulations
HCB	=	hexachlorobenzene
HCCH	=	alpha-hexachlorocyclohexane
HCCPD	=	hexachlorocyclopentadiene
HCFC	=	hydrochlorofluorocarbon
HCl	=	hydrochloric acid
HCP	=	hypothermal coal process
HCS	=	hazard communication standard
HDD	=	(1) heavy duty diesel; (2) chlorinated dibenzo-*p*-dioxin
HDF	=	dibenzofurans
HDPE	=	high-density polyethylene
HDT	=	highest dose tested in a study
He	=	helium
HEA	=	(1) health effects advisory; (2) health effects assessment
HELP	=	hazard evaluation of leakage potential
HEM	=	human exposure modeling
Hf	=	hafnium
Hg	=	mercury
HHV	=	higher heating value
HHW	=	household hazardous waste
HI	=	(1) Hydraulic Institute; (2) hazard index
HI-VOL	=	high-volume sampler
HIWAY	=	a line source model for gaseous pollutants
HLRW	=	high-level radioactive waste
HMBP	=	Hazardous Materials Business Plans, H&S Code 25504
HMI	=	Hoist Manufacturers Institute
HMR	=	hazardous materials regulations
HMTA	=	Federal Hazardous Materials Transportation Act
HMTUSA	=	Hazardous Materials Transportation Uniform Safety Act
HNu™	=	photoionization detector
Ho	=	holmium
HOC	=	(1) halogenated organic carbons; (2) hazardous organic constituents
HOMO	=	homogenous (well log)
HON	=	Hazardous Organic National Emission Standard for Hazardous Air Pollutants
hp	=	horsepower
HPC	=	heterotrophic bacteria
HPLC	=	high-performance liquid chromatography
HPLC/fluo.	=	high-performance liquid chromatography analysis with fluorescence detector

HPLC/PB/MS	=	high-performance liquid chromatography/particle beam/mass spectrometry
HPLC/TSP/MS	=	high-performance liquid chromatography/thermospray/mass spectrometry
HPLC/UV	=	high-performance liquid chromatography analysis with UV detector
HPLC-k'	=	HPLC capacity factor correlation
HPLC-RI	=	HPLC retention index correlation
HPLC-RT	=	HPLC retention time correlation
HPLC-RV	=	HPLC retention volume correlation
HPO	=	hydrous pyrolysis/oxidation
HPV	=	high-priority violators
HQ	=	hazard quotient
hr	=	hour
HRS	=	hazard ranking system (for NPL)
H&SC	=	Health and Safety Code
HSCD	=	hazardous substance database
HSD	=	halogen-specific detector
HSL	=	hazardous substance list
HSWA	=	Federal Hazardous and Solid Waste Amendments
HT	=	hypothermally treated
HTP	=	high temperature and pressure
HUD	=	Department of Housing and Urban Development
HV	=	hydrocarbon vapor
HVOC	=	halogenated volatile organic compounds
HW	=	hazardous waste
HWCA	=	Hazardous Waste Control Act
HWERL	=	hazardous waste engineering research laboratory
HWGTF	=	hazardous waste groundwater task force
HWM	=	hazardous waste management
HWRTF	=	hazardous waste restrictions task force
HWS	=	hazardous waste site
HWTC	=	Hazardous Waste Treatment Council

I

I	=	(1) iodine; (2) hydraulic gradient
IAMPO	=	International Association of Plumbing and Mechanical Officials
IAP	=	indoor air pollution
IARC	=	International Agency for Research on Cancer
IATDB	=	interim air toxics database

IBT	=	Industrial Biotest Laboratory
IC	=	integrated circuit
I&C	=	instrumentation and control
ICAP	=	inductively coupled argon plasma
ICB	=	information collection budget
ICBO	=	International Conference of Building Officials
ICE	=	(1) industrial combustion emissions model; (2) internal combustion engine
ICP	=	inductively coupled plasma spectroscopy
ICP/MS	=	inductively coupled plasma/mass spectroscopy
ICP-OES	=	inductively coupled argon plasma optical emissions spectrometer
ICRE	=	ignitability, corrosivity, reactivity, extraction
ICRP	=	International Committee on Radiation Protection
ICWM	=	Institute for Chemical Waste Management
I.D.	=	inside diameter
IDL	=	instrument detection limit
IEEE	=	Institute of Electrical and Electronic Engineers
IERL	=	Industrial Environmental Research Laboratory
IES	=	Illuminating Engineering Society
IFR	=	internal floating roof
IGN	=	igneous (well log)
IGTS	=	Interim Groundwater Treatment System
IGWMC	=	International Groundwater Modeling Center (Holcomb Research Institute)
IMPACT	=	Integrated Model of Plumes and Atmosphere in Complex Terrain
IMS	=	ion mobility spectrometry
in	=	inch
In	=	indium
in^2	=	square inch
in^3	=	cubic inch
IND	=	individual (well log)
INDURD	=	indurated (well log)
INES	=	international nuclear event scale
INP	=	initial notification plan
INPUFF	=	Gaussian puff dispersion model
INTEDD	=	interbedded (well log)
IOB	=	iron ore beneficiation
IOC	=	inorganic chemicals
IP	=	(1) implementation plan; (2) inhalable particulates; (3) ionization potential
IPCE	=	International Power Cable Engineers Association

Ir	=	iridium
IR	=	(1) infrared (light wavelength); (2) infrared spectroscopy; (3) iron (well log)
IRA	=	interim response action
IRIS	=	Integrated Risk Information System
IRM	=	interim remedial measures
IROX	=	iron oxide (well log)
IRP	=	installation of restoration program
IRPTC	=	International Register of Potentially Toxic Chemicals
IRR	=	Institute of Resource Recovery
IS	=	interim status
ISA	=	Instrument Society of America
ISCLT	=	industrial source complex long-term model
ISCST	=	industrial source complex short-term model
ISD	=	interim status document
ISE	=	ion specific electrode
ISMAP	=	indirect source model for air pollution
IWC	−	in-stream waste concentration
IWS	=	ionizing wet scrubber

J

J	=	(1) joule; (2) intermediate quantities for fugacity calculation
JAPCA	=	*Journal of Air Pollution Control Association*
JPA	=	joint permitting agreement
JTU	=	Jackson turbidity unit

K

k	=	(1) intrinsic permeability (sometimes designated by "p"); (2) first-order rate constant (hr^{-1})
K	=	(1) coefficient of permeability; (2) hydraulic conductivity of a porous medium; (3) potassium (kalium); (4) degrees Kelvin
K_2	=	elimination/clearance/depuration rate constant (d^{-1})
K_A	=	air-water mass transfer coefficient, air-side (m/hr)
K_{AW}	=	dimensionless air/water partition coefficient
K_b	=	biodegradation rate constant (d^{-1})
K_B	=	bioconcentration factor
K_h	=	(1) association coefficient; (2) hydrolysis rate constant (d^{-1})
K_H	=	Henry's Law constant

K_i	=	first-order rate constant in phase i (hr^{-1})
K_l	=	uptake/accumulation rate constant (d^{-1})
K_O	=	effective hydraulic conductivity
K_{OC}	=	organic carbon partition coefficient
K_{OM}	=	organic-matter sorption partition coefficient
K_{OW}	=	octanol-water partition coefficient
K_p	=	photolysis rate constant (d^{-1})
K_P	=	sorption coefficient
K_{sp}	=	plant/soil partition coefficient
K_W	=	air-water mass transfer coefficient, water-side (m/hr)
KEMOD	=	kinetic and equilibrium model
kg	=	kilogram
kg/L	=	kilogram per liter
km	=	kilometer
km^2	=	square kilometer
km^3	=	cubic kilometer
kPa	=	kilopascals
Kr	=	krypton
kW	=	kilowatt
kWh	=	kilowatt hour

L

L	=	lipid content of fish
La	=	lanthanum
LADD	=	(1) lifetime average daily dose; (2) lowest acceptable daily dose
LADI	=	lifetime allowable daily intake
LAER	=	lowest achievable emissions rates (Clean Air Act)
LAI	=	laboratory audit inspection
lb	=	pound
lb/mmscf	=	pounds per million standard cubic feet
LC	=	(1) lethal concentration; (2) liquid chromatography
LC_{50}	=	concentration of a toxicant that is lethal to 50% of the tested organisms
LCD	=	(1) local climatological data; (2) liquid crystal display
LCL	=	lower control limit
LCS	=	laboratory control sample
LD_{10}	=	lowest dosage of a toxic substance that kills test organisms
LD_{50}	=	dose of a toxicant that is lethal to 50% of the organisms tested in a specified time under specified test conditions
LDAR	=	leak detection and repair

LDD	=	light duty diesel
LDR	=	land disposal restriction
LDS	=	leak detection system
LDT	=	lowest dose tested
LEL	=	(1) lower explosive limit; (2) lowest effect level
LEPC	=	local emergency planning committee
LERC	=	local emergency response committee
LET	=	linear energy transfer
LFL	=	lower flammability limit
L/hr/m^2	=	liters per hour per square meter
LHW	=	liquid hazardous waste
LI	=	liquidity index
Li	=	lithium
LIF	=	laser-induced fluorescence system
LIMB	=	limestone injection multi-stage burner
LIMS	=	laboratory information management system
LL	=	liquid limit (%)
LLDPE	=	linear low-density polyethylene
LLNL	=	Lawrence Livermore National Laboratory
LLRW	=	low-level radioactive waste
LMMTD	=	laminated (well log)
LMV	=	light- and medium-weight vehicles
LOD	=	limit of detection
LOE	=	level of effort
LOEL	=	lowest observed effect level
LOI	=	letter of intent
LOIS	=	loss of interim status
LOQ	=	limit of quantitation
lpm	=	liters per minute
Lr	=	lawrencium
LRT	=	liquid release test
ls	=	limestone
LS	=	loose (well log)
LSC	=	liquid scintillation counting
LT	=	light (well log)
LTDL	=	lower than detection limits
LTEL	=	long-term exposure limit
LTL	=	little (well log)
Lu	=	lutetium
LUFT	=	leaking underground fuel tank
LUST	=	leaking underground storage tank
LVF	=	liquid-volume fraction
LVS	=	laboratory vane shear

M

μ	=	dynamic viscosity
μg	=	microgram; 1/1000 of a milligram
$\mu g/g$	=	micrograms/gram
$\mu g/L$	=	micrograms per liter (10^{-6})
$\mu g/m^3$	=	micrograms per cubic meter
$\mu g/mL$	=	microgram/milliliter
μL	=	microliter; 1/1,000,000 of a liter
μsec	=	microsecond
m	=	meter
m/day	=	meters per day
m/min	=	meters per minute
m^2	=	square meter
m^3	=	cubic meter
m^3/day	=	cubic meters per day
m^3/sec	=	cubic meters per second
m_i	=	amount of chemical in phase i (mol or kg)
M	=	(1) medium/moderate (well log); (2) total amount of chemical (mol or kg)
MA	=	particle size analysis
MAC	=	maximum allowable concentration
MACT	=	maximum achievable control technology
MAPSIM	=	mesoscale air pollution simulation model
mb	=	millibars
MBAS	=	methylene-blue-active substances
MBK	=	methyl n-buytl ketone
MCA	=	mica/micaceous (well log)
MCH	=	methylcyclohexane
MCL	=	maximum contaminant level (see RMCL)
MCLG	=	maximum contaminant level goals
Md	=	mendelevium
MDA	=	methylene disalicyclic acid
MDD	=	maximum daily dose
MDL	=	method detection limit
ME	=	micro-extraction
MEBK	=	methyl ethyl butyl ketone
MEK	=	methyl ethyl ketone
MEL	=	maximum exposure limits
MeOH	=	methanol
MEOR	=	microbial enhanced oil recovery
MEP	=	multiple extraction procedure
MFS	=	minimum functional standards

mg	=	milligram
Mg	=	magnesium
mg/dscm	=	milligrams per dry standard cubic meter
mg/kg	=	milligram/kilogram
mg/L	=	milligrams per liter (10^{-3})
Mg/yr	=	megagrams per year
MGD	=	millions of gallons per day
MH	=	inorganic silts, micaceous or diatomaceous fine sandy or silty soils, elastic silts (well log)
MHT	=	maximum holding time
MH_z	=	megahertz
mi	=	mile
mi^2	=	square mile
mi^3	=	cubic mile
MIBK	=	methyl isobutyl ketone
MJR	=	major (well log)
ML	=	inorganic silts and very fine sands
mL	=	milliliter
mm	=	millimeter
mm^2	=	square millimeter
mm^3	=	cubic millimeter
mmL/yr	=	million liters per year
mmscf	=	million standard cubic feet
MMT	=	methylcyclopentadienyl manganese tricarbonyl
mn	=	manganese
MNOX	=	manganese oxide (well log)
MNR	=	minor (well log)
MO	=	molecular orbital calculation
Mo	=	molybdenum
MOA	=	memorandum of agreement
MOU	=	memorandum of understanding
MP	=	measuring point (usually assigned to a monitoring well)
M.P.	=	melting point, °C
MPRSA	=	Marine Protection Resource and Sanctuaries Act
MR	=	molar refraction
MRL	=	maximum-residue limit (pesticide tolerance)
ms	=	millisecond
MS	=	mass spectrometry
MSDS	=	Material Safety Data Sheets
MSL	=	mean sea level
MSSV	=	massive (well log)
MST	=	moist (well log)
MTBE	=	methyl tertiary butyl ether

MTTLING	=	mottling (well log)
MTU	=	mobile treatment unit
MW	=	molecular weight (g/mol)
MWC	=	municipal waste combuster
MWL	=	municipal waste leachate

N

v	=	kinematic viscosity (mm^2/sec)
n	=	porosity
n_C	=	number of carbon atoms
n_{Cl}	=	number of chlorine atoms
n_e	=	effective porosity
n_p	=	coefficient of permeability
N	=	(1) nitrogen; (2) no, none, non-
Na	=	sodium (natrium)
NA	=	not analyzed, not available, not applicable
NAA	=	(1) non-attainment area; (2) neutron activation analysis
NAAQS	=	National Ambient Air Quality Standards (Clean Air Act)
NAC	=	National Asbestos Council
NAEP	=	National Association of Environmental Professionals
NANCO	=	National Association of Noise Control Officials
NAPL	=	non-aqueous phase liquids
NAR	=	National Asbestos Registry
NARS	=	National Asbestos-Contractor Registry System
NAS	=	National Academy of Science
NAWC	=	National Association of Water Companies
Nb	=	niobium
NBARs	=	non-binding preliminary allocations of responsibility (used in conjunction with CERCLA investigations)
NBS	=	National Bureau of Standards
NCAPAH	=	non-carcinogenic polycyclic aromatic hydrocarbons
NCP	=	National (Oil and Hazardous Substances Pollution) Contingency Plan
NCS	=	notification of compliance status
NCWQ	=	National Commission of Water Quality
Nd	=	neodymium
ND	=	non-detected
NDD	=	negotiated decision document
Ne	=	neon
NEC	=	National Electric Code
NEIC	=	National Enforcement Investigations Center

NEMA	=	National Electric Manufacturers Association
NEPA	=	National Environmental Policy Act
NERL-CRD	=	National Exposure Research Laboratory — Characterization Research Division
NESHAP	=	National Emission Standards for Hazardous Air Pollutants
NFPA	=	National Fire Protection Association
NFRAP	=	no further remedial action planned
NFWE	=	no free water encountered
NGA	=	National Governors Association
NGWIX	=	National Groundwater Information Center
NH_3	=	ammonia
NH_3-N	=	ammonia nitrogen
Ni	=	nickel
NICT	=	National Incident Coordination Team
NIMBY	=	"not in my backyard"
NIOSH	=	National Institute for Occupational Safety and Health
NIPDWR	=	National Interim Primary Drinking Water
nm	=	nanometer
NMOC	=	non-methane organic compound
NMR	=	nuclear magnetic resonance
No	=	nobelium
NOAA	=	National Oceanic and Atmospheric Administration
NOAEL	=	no observable adverse effect level
NOC	=	nonpolar, non-ionic organic compound
NOD	=	notice of deficiency (permit response)
NOEL	=	no observable effect level
NOI	=	notice of intent
NON	=	notice of noncompliance
NOSC	=	no odor, scent, or fluid cut
NOV	=	notice of violation
NO_x	=	nitrous oxides
Np	=	neptunium
NPAA	=	Noise Pollution and Abatement Act
NPD	=	nitrogen phosphorous detector
NPDES	=	National Pollutant Discharge Elimination System
NPDWR	=	national primary drinking water regulations
NPHAP	=	National Pesticide Hazardous Assessment Program
NPL	=	National Priorities List
NPLST	=	non-plastic (well log)
NPO	=	no product odor (well log)
NPR	=	notice of proposed rulemaking
NRC	=	Nuclear Regulatory Commission
NRDC	=	Natural Resources Defense Council

NRT	=	National Response Team
NS	=	nonspecified
NSDWR	=	National Secondary Drinking Water Regulations
NSF	=	National Sanitation Foundation
NSPS	=	New Source Performance Standards (Clean Water Act)
NTIS	=	National Technical Information Service
NTWS	=	non-transient/non-community water systems
NWA	=	National Water Alliance
NWF	=	National Wildlife Federation
NWPA	=	Nuclear Waste Policy Act
NWWA	=	National Water Well Association

O

O	=	odor (well log)
O_2	=	oxygen
OADEMQA	=	Office of Acid Deposition, Environmental Monitoring, and Quality Assurance (EPA)
OCPSF	=	organic chemicals, plastics, and synthetic fibers
OCS	=	outer continental shelf
O.D.	=	outside diameter
OEA	=	Office of Environmental Affairs (California)
OECM	=	Office of Enforcement and Compliance Monitoring
OER	=	Office of Exploratory Research (EPA)
OERR	=	Office of Emergency and Remedial Response (EPA)
OH	=	organic clays of medium to high plasticity, organic silty clays (well log)
OHR	=	Office of Health Research (EPA)
OILHM	=	Oil and Hazardous Material Information System
OL	=	organic silts and organic silty clays
O&M	=	operation and maintenance
OMB	=	Office of Management and Budget
ORAD	=	Operations and Regulatory Affairs Division
ORC	=	Office of Regional Counsel
ORD	=	Office of Research and Development (EPA)
ORG	=	organic (well log)
ORNG	=	orange (well log)
ORNL	=	Oak Ridge National Laboratory
ORP	=	oxidation-reduction potential
ORPM	=	Office of Research Program Management (EPA)
Os	=	osmium
OSC	=	on-scene coordinator

OSHA	=	(1) Occupational Safety and Health Act; (2) Occupational Safety and Health Administration
OSM	=	Office of Surface Mining
OSS	=	site safety officer
OSWER	=	Office of Solid Waste and Emergency Response (EPA)
OTA	=	Office of Technology Assessment
OVA	=	organic vapor analyzer
OVM	=	organic vapor meter
oz	=	ounce

P

p_o	=	atmospheric pressure
P	=	(1) phosphorous; (2) vapor pressure (Pa)
P_L	=	liquid or subcooled liquid vapor pressure (Pa)
P_S	=	solid vapor pressure (Pa)
Pa	=	(1) protactinium; (2) Pascal
PA	=	preliminary assessment
PAH	=	polynuclear aromatic hydrocarbon (also PNA)
PAI	=	(1) performance audit inspection (Clean Water Act); (2) pure active ingredient
PAIR	=	preliminary assessment information rule
PAM	=	*Pesticide Analytical Manual*
PAN	=	peroxyacetylnitrate
PARCON	=	plasma arcing conversing
PAT	=	plasma arc technology
PATS	=	pesticide analytical transport solution
Pb	=	lead
PBB	=	polybrominated biphenyl
PCA	=	principal component analysis
PCB	=	polychlorinated biphenyl
PCE	=	tetrachloroethlyene, also perchloroethylene and perc
pCi/L	=	picocuries per liter
PCM	=	phase contrast microscopy
PCP	=	pentachlorophenol
Pd	=	palladium
PDA	=	photodiode array
PDE	=	pre-design evaluation
PDQ	=	practical quantitation limit
P.E.	=	Professional Engineer
PEL	=	(1) permissible exposure limit; (2) polyester elastometer
PER	=	perchloroethylene (tetrachloroethylene)

Perm	=	permeability
PET	=	polyethylene terephthalate
PFRP	=	procedure to further enhance pathogens
P.H.	=	Professional Hydrogeologist
pH	=	value of acidity and alkalinity
PHA	=	public health assessment
PHC	=	principal hazardous constituent
PHE	=	public health evaluation
PHERE	=	public health environmental risk evaluation
PHNO	=	phenocryst (well log)
PI	=	plasticity index
PIC	=	products of incomplete combustion
PID	=	photoionization detector
PIG	=	program implementation guide
PIGS	=	pesticides in groundwater strategy
PIN	=	procurement information notice
PLASM	=	Prickett-Lonnquist aquifer simulation model
PLCC	=	plastic leaded chip carrier
PLIRRA	=	Pollution Liability Insurance and Risk Retention Act
PLM	=	polarized light microscopy
PLST	=	plastic/plasticity (well log)
PLUVUE	=	plume visibility model
Pm	=	promethium
PMN	=	pre-manufacture notice
PMT	=	photomultipler tube
PNP	=	p-nitrophenol
Po	=	polonium
POC	=	purgeable organic carbon
POE	=	point of entry
POG	=	point of generation
POHC	=	principal organic hazardous constituent
POI	=	point of interception
POM	=	(1) particulate organic matter; (2) polycyclic organic matter
POSS	=	possible (well log)
POTWs	=	publicly owned treatment works
ppb	=	parts per billion (μg/L or μg/kg)
PPE	=	personal protective equipment
PPIC	=	Pesticide Programs Information Center
PPL	=	priority pollutant list
ppm	=	parts per million (mg/L or mg/kg)
ppmv	=	parts per million by volume
ppt	=	(1) parts per trillion; (2) parts per thousand; (3) precipitate
Pr	=	praseodymium

PRA	=	Paperwork Reduction Act
PRAP	=	proposed remedial action plan
PRATS	=	Pesticides Regulatory Action Tracking System
PRLY	=	poorly, product (well log)
PRP	=	potentially responsible party (CERCLA)
PRPH	=	porphyry/porphyritic (well log)
PRZM	=	pesticide root zone model
PSA	=	preliminary site assessment
PSAM	=	point source ambient monitoring
PSD	=	Prevention of Significant Deterioration (Clean Air Act air permit)
psi	=	pounds per square inch
psia	=	pounds per square inch atmosphere
psig	=	pounds per square inch at gauge
Pt	=	platinum
PT	=	(1) purge and trap; (2) peat and other highly organic soils
PTA	=	packed tower aeration
PTDIS	=	single stack meteorological model
PTFE	=	polytetrafluoroethylene, Teflon®
PTPLU	=	point source Gaussian diffusion model
Pu	=	plutonium
PVA	=	polytopic vector analysis
PVC	=	polyvinyl chloride (typical well-construction material)
PWS	=	public water system
PWSS	=	public water supply system

Q

q	=	special discharge, or special flux
Q	=	(1) discharge rate (i.e., rate of groundwater discharge from a pumping well) or total flux; (2) scavenge ratio
QA	=	quality assurance
QA/QC	=	quality assurance/quality control
QAMS	=	quality assurance management staff
QAO	=	quality assurance officer
QAPP	=	quality assurance project plan
qBtu	=	quadrillion British thermal units
QC	=	quality control
QIP	=	quality improvement plan
qs	=	vane-sheer strength
QSAR	=	quantitative structural-activity relationship
QSPR	=	quantitative structure-property relationship
qt	=	quart

QTI	=	quartz (well log)
qtz	=	quartzite
qu	=	unconfined compressive strength
QUIPE	=	*Quarterly Update for Inspectors in Pesticide Enforcement*
QW	=	quality of water

R

ρ	=	density (g/cm^3)
R	=	rapid (well log)
R_e	=	effective porosity
Ra	=	radium
RA	=	(1) risk assessment; (2) risk analysis; (3) regional administrator (EPA)
RAA	=	remedial action alternative
RAATS	=	RCRA Administrative Action Tracking System
RACT	=	reasonably available control technology (Clean Air Act)
RAD	=	(1) radiation; (2) radiation adsorbed dose
RADM	=	(1) random walk advection and dispersion model (groundwater); (2) regional acid deposition model
RAMP	=	(1) remedial action master plan; (2) Rural Abandoned Mine Program
RAP	=	(1) Remedial Action Plan (Program); (2) Radon Action Program; (3) response action plan
RAT	=	remedial action technology
Rb	=	rubidium
RB	=	rock-roller bit
RCPP	=	Revised RCRA Civil Penalty Policy
RCRA	=	Resource Conservation and Recovery Act
RCRIS	=	Resource Conservation and Recovery Information System
RCT	=	reference control technology
RD	=	(1) red (well log); (2) remedial design
RD/RA	=	remedial design/remedial action
RDF	=	refuse-derived fuel
RDV	=	reference dose value
R.E.	=	Registered Environmental Assessor
R.E.A.	=	Registered Environmental Assessor (California)
REM	=	Roentgen equivalent in man
REM/FIT	=	Remedial Planning and Field Investigation Team (contracted to EPA)
RFA	=	RCRA facility assessments
RfD	=	risk reference doses

RFI	=	remedial facility investigation
RFID	=	radio frequency identification
R.G.	=	Registered Geologist
RH	=	relative humidity
Rh	=	rhodium
RHRS	=	Revised Hazard Ranking System
RHW	=	restricted hazardous waste
RI	=	remedial investigation
RI/FS	=	Remedial Investigation and Feasibility Study
RIA	=	radio immunoassay
RIC	=	remote intelligent communications
RIM	=	regulatory interpretive memorandum
RKS	=	rocks (well log)
RMCL	=	recommended maximum contaminant level
RMDHS	=	Regional Model Data Handling System
RMPP	=	Risk Management and Prevention Program
Rn	=	radon
RNA	=	ribonucleic acid
RND	=	round (well log)
ROADWAY	=	model to predict pollutant concentrations near a roadway
ROD	=	record of decision
ROST	=	rapid optical screen tool
RP	=	responsible parties
RPD	=	relative percent difference
RP-HPLC	=	reversed phase, high-pressure liquid chromatography
RPM	=	(1) reactive plume model; (2) remedial project manager (EPA)
rpm	=	revolutions per minute
rps	=	revolutions per second
RP-TLC	=	reversed phase, thin-layer chromatography
RQ	=	reportable quantity
RSD	=	(1) relative standard deviation; (2) risk-specific dose
RSPA	=	Research and Special Programs Administration (DOT)
RTECS	=	Registry of Toxic Effects of Chemical Substances
RTM	=	regional transport model
Ru	=	ruthenium
RVP	=	Reid vapor pressure
RX	=	reactive (well log)

S

s	=	drawdown
S	=	(1) storage coefficient or storativity; (2) sulfur; (3) water solubility (mg/L or g/m^3)

ΔS_{fus}	=	entropy of fusion (J/mol • K or cal/mol • K [e.u.])
$S_{octanol}$	=	solubility in octanol
S_r	=	special retention
SA	=	sub-angular (well log)
SAB	=	Science Advisory Board
SAP	=	Scientific Advisory Panel
SAR	=	sodium adsorption ratio
SARA	=	Superfund Amendments and Reauthorization Act
SARM	=	standard analytical reference material
SAROAD	=	Storage and Retrieval of Aerometric Data
SAVE	=	spray aeration extraction system
Sc	=	scandium
SCAP	=	Superfund Comprehensive Accomplishment Plan
SCAPS	=	site characterization and analysis penetrometer system
SCBA	=	self-contained breathing apparatus
SCC	=	Source Classification Code
scfm	=	standard cubic feet per minute
sch	=	schist
scmm	=	standard cubic meters per minute
SCR	=	selective catalytic reduction
SCS	=	Soil Conservation Service
SCSA	=	Soil Conservation Society of America
SCW	=	supercritical water oxidation
SD	=	standard deviation
SDL	=	soluble designated level
SDR	=	standard dimension ratio
SDRAM	=	synchronous DRAM
SDWA	=	Federal Safe Drinking Water Act
Se	=	selenium
sec	=	second
SEC	=	Site Evaluation Submission
SEM	=	(1) scanning electronic microscope; (2) standard error of the means
SEP	=	supplemental environmental projects
SES	=	secondary emissions standard
SETS	=	Site Enforcement Tracking System
SF	=	Superfund
SFA	=	spectral flame analyzer
SFT	=	soft (well log)
SGRAM	=	synchronous graphics DRAM
SHSO	=	site health and safety officer
SHWL	=	seasonal high water level
Si	=	silicon

SI	=	(1) site investigation; (2) International Systems of Units
SIC	=	Standard Industrial Classification
SIL	=	siliceous (well log)
SIMM	=	single in-line memory module
SIMS	=	secondary ion mass spectrometry
SIP	=	Site (State) Implementation Plan (Clean Air Act)
SITE	=	Superfund Innovative Technology Evaluation Program
SL	=	(1) shrinkage limit; (2) slightly
SLPO	=	slight product odor (well log)
SLSM	=	simple line source model
slst	=	silt stone
SLT	=	silt (well log)
SLTY	=	silty (well log)
SLW	=	slow (well log)
Sm	=	samarium
SM	=	silty sands (well log)
SMACNA	=	Sheet Metal and Air Conditioning Contractors National Association
SMCL	=	secondary maximum contaminant level
SMCRA	=	Surface Mining Control and Reclamation Act
Sn	=	tin
SNAAQS	=	Secondary National Ambient Air Quality Standards
SNAP	=	Significant Noncompliance Action Program
SNARL	=	suggested no adverse response level
SND	=	sand (well log)
SNDST	=	sandstone (well log)
SNL	=	Sandia National Laboratories
SNUR	=	significant new use rules
SO_2	=	sulfur dioxide
SOC	=	(1) synthetic organic chemical; (2) suspended organic carbon
SOCMI	=	Synthetic Organic Chemical Manufacturing Industry
SOJ	=	small outline j-leaded plastic package
SOP	=	standard operating procedure
SOW	=	Statement of Work
SP	=	(1) poorly sorted soil; (2) spontaneous potential log; (3) poorly graded sands or gravelly sands
SPCC	=	spill prevention, control (containment), and countermeasures
SPDES	=	surface-water pollutant discharge elimination system
SPECS	=	specifications
SPLMD	=	soil pore liquid monitoring device
SPME	=	solid phase micro-extraction
SPOC	=	single point of contact

SPSS	=	Statistical Package for the Social Sciences
SPT	=	standard (cone) penetrometer testing
SQBE	=	small quantity burner exemption
SQG	=	small quantity generators
SR	=	(1) styrene rubber; (2) sub-round (well log)
Sr	=	strontium
SRAM	=	static random access memory
SRAP	=	Superfund Remedial Accomplishment Plan
SRC	=	solvent refined coal
SRIA	=	surface reflectance immunoassay
SRM	=	standard reference materials
SRTD	=	sorted (well log)
ss	=	sandstone
SS	=	(1) split-spoon; (2) suspended solids
SSA	=	sole source aquifer
SSAC	=	soil site assimilated capacity
SSCD	=	simple shear consolidated drained test
SSCU	=	simple shear consolidated undrained test
SSI	=	size-selective inlet
SSPC	=	Structural Steel Painting Council
ST	=	Shelby tube
STARS	=	Superfund Transactions Automated Retrieval System
STCKY	=	sticky (well log)
STELs	=	short-term exposure limits
STEM	=	scanning transmission electron microscope
STLC	=	soluble threshold limit concentration
STORET	=	storage and retrieval of water-related data
STRNG	=	strong (well log)
STRNGPO	=	strong product odor (well log)
SUD	=	safe use determination
SUP	=	standard unit of processing
SURF	=	surface (well log)
SVE	=	soil vapor extraction
SW	=	(1) solid waste; (2) well-graded soil; (3) well-graded sands or gravelly sands
SWAT	=	Solid Waste Assessment Test
SWDA	=	Federal Solid Waste Disposal Act
SWDSCMA	=	Solid Waste Disposal Site Cleanup Management Account
SWIS	=	Active and Inactive Sanitary Landfills and Disposal Facilities (State of California Waste Management Board)
SWL	=	static water level
SWMU	=	Solid Waste Management Unit
SWRCB	=	State Water Resources Control Board

SWTR	=	surface water treatment rule
S_y	=	specific yield

T

t	=	residence time (hr)
t_0	=	overall residence time (hr)
$t_{1/2}$	=	half-life (hr)
t_A	=	advection persistence time (hr)
t_B	=	sediment burial residence time (hr)
t_R	=	reaction persistence time (hr)
T	=	(1) transmissivity of an aquifer, equal to the hydraulic conductivity (K) × saturated thickness (b)/T = Kb; (2) system temperature (K)
T_B	=	boiling point (K)
T_{ij}	=	intermedia transport rate (mol/hr or kg/hr)
T_M	=	melting point (K)
T_xCD	=	consolidated drained triaxial shear test
T_xCU	=	consolidated undrained triaxial shear test
T_xUU	=	unconsolidated undrained triaxial shear test
Ta	=	tantalum
TAG	=	technology assistance grant
TALMS	=	tunable atomic line molecular spectroscopy
TAMS	=	toxic air monitoring system
TAR	=	Technical Amendment to the Regulations
Tb	=	terbium
TBC	=	to be considered
TBT	=	tributyltin
Tc	=	technetium
TC	=	toxicity characteristic
TCA	=	1,1,2-trichloroethane
TCB	=	1,2,4-trichlorobenzene
TCDD	=	tetrachlorodibenzo-*p*-dioxin (dioxin)
TCDF	=	tetrachlorodibenzofurans
TCE	=	trichloroethylene
TCFM	=	trichlorofluoromethane
TCLP	=	toxicity characteristic leaching procedure
TCP	=	2,4,6-trichlorophenol
TCRI	=	toxic chemical release inventory
TD	=	(1) toxic dose; (2) total depth (usually used in connection with wells)
TDR	=	time domain reflectometry

TDS	=	total dissolved solids
Te	=	tellurium
TEG	=	tetraethylene glycol
TEH	=	total extractable hydrocarbons
TEL	=	tetraethyl lead
TEM	=	Texas episodic model
TEML	=	triethylmethyl lead
TES	=	Technical Enforcement Support
TETS	=	totally enclosed treatment system
TFF	=	tuff/tuffaceous (well log)
TGHNSS	=	toughness (well log)
Th	=	thorium
THC	=	total hydrocarbons
THM	=	trihalomethanes
Ti	=	titanium
TIC	=	tentatively identified compound
TIS	=	Tolerance Index System
TKN	=	total Kjehldahl nitrogen
Tl	=	thallium
TLC	=	thin layer chromatography
TLV	=	threshold limit value
TLV–STEL	=	threshold limit value–short-term exposure limit
TLV–TWA	=	threshold limit value–time-weighted average
Tm	=	thulium
TMDL	=	total maximum daily limit; total maximum daily load
TMEL	=	trimethylethyl lead
TML	=	tetramethyl lead
TMP	=	2,2,4-trimethylpentane
TMP/MCH	=	2,2,4-trimethylpentane-to-methylcyclohexane ratio
TMRC	=	theoretical maximum residue contribution
TMV	=	total molecular volume per molecule (angstrom^3)
TNT	=	trinitrotoluene
TOA	=	trace organic analysis
TOC	=	(1) total organic compound; (2) total organic carbon
TOG	=	total oil and grease
TOH	=	total organic halogens (*see also* TOX)
TOT	=	time of travel
TOX	=	(1) tetradichloroxylene; (2) total organic halogens (*see* TOC)
TPCA	=	Toxic Pits Cleanup Act H&S Code 25208 et seq.
TPH	=	total petroleum hydrocarbons
TPTH	=	triphenyltinhydroxide
TPY	=	tons per year
TR	=	trace (well log)

T-R	=	transformer-rectifier
trans-DCE	=	*trans*-dichloroethylene
TRE	=	toxic release effectiveness index
TRI	=	toxic release inventory
TRIP	=	Toxic Release Inventory Program
TRIS	=	Toxic Release Inventory System
TRPH	=	total recoverable petroleum hydrocarbons
TS	=	total solids
TSA	=	total surface area per molecule
TSCA	=	Toxic Substances Control Act
TSD	=	treatment, storage, and disposal
TSDF	=	treatment, storage or disposal facility (hazardous waste)
TSI	=	thermal system insulation
TSOP	=	thin small outline package
TSP	=	(1) tri-sodium phosphate; (2) total suspended particulates
TSS	=	total suspended solids
TTFA	=	target transformation factor analysis
TTHM	=	total trihalomethane
TTLC	=	total threshold limit concentration
TTO	=	total toxic organics
TVH	=	total volatile hydrocarbons
TWA	=	time-weighted average

U

U	=	uranium
U_1	=	air-side, air/water MTC (same as k_A) (m/hr)
U_2	=	water-side, air/water MTC (same as k_A) (m/hr)
U_3	=	rain rate (same as U_R) (m/hr)
U_4	=	aerosol deposition rate (m/hr)
U_5	=	soil/air phase diffusion MTC (m/hr)
U_6	=	soil/water phase diffusion MTC (m/hr)
U_7	=	soil/air boundary layer MTC (m/hr)
U_8	=	sediment-water MTC (m/hr)
U_9	=	sediment deposition rate (m/hr)
U_{10}	=	sediment resuspension rate (m/hr)
U_{11}	=	soil-water runoff rate (m/hr)
U_{12}	=	soil-solids runoff rate (m/hr)
U_B	=	sediment burial rate (m/hr)
U_Q	=	dry deposition velocity (m/hr)
U_R	=	rain rate (m/hr)
U&A	=	understanding and agreement

UAM	=	urban airshed model
UBC	=	Uniform Building Code
UC	=	unconfined compression
UCC	=	ultra-clean coal
UCCI	=	urea-formaldehyde foam insulation
UCL	=	upper control limit
UCR	=	unit carcinogenic risk
UDMH	=	unsymmetrical dimethyl hydrazine
UEL	=	upper explosive limit
UFC	=	Uniform Fire Code
UFL	=	upper flammability limit
UGST	=	underground storage tank
UGT	=	underground tank
UIC	=	underground injection control
UL	=	Underwriters Laboratories
UM	=	uniform manifest
UMTRCA	=	Uranium Mill Tailings Radiation Control Act
UNIFAC	=	UNIQUAC Functional Group Activity Coefficients
Unp	=	unnilpentium
Unq	=	unnilquadium
USAMBRDL	=	U.S. Army Medical and Biological Research and Development Laboratory
USATHAMA	=	U.S. Army Toxic and Hazardous Materials Laboratory
USC	=	(1) Unified Soil Classification; (2) United States Code
USCOE	=	U.S. Army Corps of Engineers (also ACE, COE, CORPS, USACE)
USCS	=	Unified Soil Classification System
USDA	=	U.S. Department of Agriculture
USDW	=	underground source of drinking water
USEPA	=	U.S. Environmental Protection Agency
USFS	=	U.S. Forest Service
USGS	=	U.S. Geological Survey (see also GS)
UST	=	underground storage tank
UTM	=	universal transverse mercator
UV	=	(1) ultraviolet; (2) UV spectrometry
UVFL	=	ultraviolet/fluorescence spectrophotometer
UV–VIS	=	ultraviolet–visible spectroscopy

V

V	=	(1) vanadium; (2) very (well log)
V_{Bi}	=	volume of bulk phase (m^3)

VCM	=	vinyl chloride monomer
VEFT	=	vacuum extraction feasibility testing
VES	=	vapor extraction system
VHS	=	vertical and horizontal spread model
v_i	=	volume fraction of phase i (m^3)
v_Q	=	volume fraction of aerosol (m^3)
V_i	=	volume of pure phase i (m^3)
V_I	=	intrinsic molar volume (m^3)
V_M	=	molar volume (m^3)
V_S	=	volume of bottom sediment (m^3)
VOA	=	volatile organic analysis
VOC	=	volatile organic compound
VOHAP	=	volatile organic hazardous air pollutant
VOL	=	volatile organic liquid
VOLC	=	volcanic (well log)
VP	=	vapor pressure
VSD	=	virtually safe dose
VSI	=	visual site inspection
VVHAP	=	very volatile hazardous air pollutant

W

w	=	moisture content
W	=	(1) tungsten (Wolfram); (2) molecular mass (g/mol)
W/	=	with (well log)
WAP	=	waste analysis plan
WB	=	wet bulb
WDR	=	waste discharge requirement
WERL	=	Water Engineering Research Laboratory
Whp	=	water horsepower
WHP	=	Wellhead Protection Program
WHPA	=	wellhead protection area
WHT	=	white (well log)
WL	=	(1) water level; (2) working level
WPCF	=	Water Pollution Control Federation
WQCB	=	Water Quality Control Board (Regional)
WQS	=	Water Quality Standards
WRCB	=	Water Resource Control Board
WRD	=	Water Resources Division (usually in connection with the USGS)
WRDA	=	Water Resources Development Act
WRDROP	=	Distribution Register of Organic Pollutants in Water

Ws	=	wash sample
WSTB	=	Wastewater Sewage Treatment Plant
WT	=	water table (well log)
WTM	=	wavelength time matrix
WTR	=	water (well log)
WWTP	=	wastewater treatment plant

X

| Xe | = | xenon |
| XRF | = | X-ray fluorescence spectrometry |

Y

Y	=	yttrium
Yb	=	ytterbium
YLLW	=	yellow (well log)

Z

Z_{Bi}	=	fugacity capacity of bulk phase i (mol/m^3 Pa)
Z_i	=	fugacity capacity of phase i (mol/m^3 Pa)
ZHE	=	zero headspace extractor
Zn	=	zinc
ZOI	=	zone of incorporation
Zr	=	zirconium
ZRL	=	zero risk level

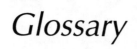

Glossary

Glossary

A

A. Abbreviation for angstrom, which is equal to 10^{-9} meters.

A-, An-. Chemical prefix meaning "not" or "without".

Abiotic. Not biotic; not formed by biologic processes.

Above-ground tank. Tank situated so that the entire surface area of the tank (including the bottom surface) is completely above the plane of the adjacent surrounding surface.

Abrasive. Substance used in particulate or bonded form to polish, smooth, or reduce the surface of another substance that is of lower hardness.

Abrasive blasting. The process for cleaning surfaces by means of such materials as sand, alumina, or steel grit in a stream of high-pressure air.

Absolute error. Difference between the true value of a parameter and the computed or measured parameter value.

Absorption. The process by which ions or molecules present in one phase tend to penetrate into and concentrate in the interior of a solid or a liquid. The reversal of absorption can be slow because the absorbed chemical must break loose and then migrate through the matrix before it can escape.

Acaricide. Class of pesticides used to kill mites and ticks; also known as "miticide".

Accuracy. Measure of the closeness of a set of measurements to the true or expected value; high accuracy implies high precision.

Acetate. An ester or salt of acetic acid in which a terminal hydrogen atom of the acid is replaced by either a hydrocarbon or other radical or metal.

Acetone. Colorless liquid with a fragrant, mint-like odor.

Acicular. Having the shape of a needle.

Acid. A liquid with a high concentration of hydrogen ion; alternatively, a substance that donates hydrogen ions. Characterized by a low pH (less than 7), acid dissolves many metals and promotes the movement of cationic metal ions in groundwater systems.

Acid igneous rock. Rock whose composition contains more than 65% silica.

Acre-foot (AF). The quantity of water that will cover an acre of land to a depth of 1 foot (i.e., 43,560 cubic feet or 325,900 gallons).

Acrylamide. Colorless, odorless crystals with a melting point of 84.5°C which are soluble in water, alcohol, and acetone.

Actinomycetes. Any of numerous, generally filamentous and often pathogenic, microorganisms resembling both bacteria and fungi.

Activated carbon. Porous form of carbon made by destructively distilling carbon-rich materials to eliminate volatile components followed by high-temperature treatment with steam or carbon dioxide.

Activated carbon adsorption. Physical treatment technology whereby soluble substances are collected on the surface of activated carbon by surface attraction phenomena.

Activated silica. A negatively charged colloidal particle formed by the reaction of a dilute sodium silicate solution with a dilute solution of an acidic material.

Activated sludge. Sludge floc produced in raw or settled wastewater by the growth of bacteria and other organisms in the presence of dissolved oxygen and accumulated in sufficient concentration by returning floc previously formed.

Activated solids. Combination of organisms in wastewater solids produced in the presence of dissolved oxygen in activated sludge treatment.

Activation analysis. Analytical technique that induces radioactivity for the purpose of quantitative analysis of elements.

Activator. Substance that accelerates the effect or increases the total effect of a pesticide

Active fault. A fault that has had surface rupture within Holocene time (the past 11,000 years).

Active ingredient. In a pesticide product, the component which kills, or otherwise controls, the target and pests. Pesticides are regulated primarily on the basis of their active ingredients.

Activity. (1) The tendency of a metal with a high electromotive value to replace another metal that is lower in the series; (2) the tendency of a metal to accelerate the chemical combination of other substances (i.e., catalytic activity).

Activity coefficient. Ratio of the effective contribution of a substance to the actual contribution of the substance to the phenomena.

Acute. Of short duration and/or rapid onset; develops during or shortly after a brief exposure to a toxic substance.

Acute effects. Symptoms that are manifested shortly after exposure.

Acute exposure. Exposure over a short period of time.

Acute toxicity. Toxic symptoms that develop shortly after exposure, usually within 24 hours.

Additive effect. Effect of a mixture which is equal to the sum of the effects of its individual components.

Adhesive. Substance that has the ability to attach itself to the surface of other substances to form a strong and relatively permanent bond.

Adhesive water. Film of water left around each grain of water-bearing material after gravity water has been drained off.

Adiabatic. Term used by chemical engineers to characterize a process in which no heat is gained or lost as the conditions of operation are changed.

Adjuvant. Used in formulation to aid the operation or improve the effectiveness of a pesticide. The term includes such materials as wetting agents, spreaders, emulsifiers, dispersing agents, foam adjuvants, foam suppressants, penetrants, and correctives. A spray adjuvant may contain one or more surfactants, solvents, solubilizers, buffering agents, and stickers necessary to formulate a specific type of adjuvant. By using the proper adjuvant, it is often possible to combine certain chemical pesticides in a tank mixture that otherwise would present compatibility problems.

Adsorbate. Relating to the adsorption process — the material being concentrated is termed the "adsorbate".

Adsorbent. Relating to the adsorption process — the adsorbing solid is termed an "adsorbent".

Adsorption. Assimilation of gas, vapor, or dissolved matter by the surface of a solid. It is a process whereby atoms, molecules, or ions are taken up and retained on the surfaces of solids by chemical or physical binding and is usually reversible. Three general types of adsorption are physical, chemical, and exchange adsorption. The reversal of adsorption requires only the breaking of the attachment before the adsorbed chemical can escape.

Advection. The process by which solutes are transported by the bulk motion of groundwater.

Advective transport. The process by which solutes are transported by the bulk motion of flowing groundwater or surface water.

Aerated lagoons. Natural or artificial wastewater ponds or basins in which mechanical or diffused air is used to supplement natural oxygen supply.

Aeration. (1) The bringing about of intimate contact between air and a liquid by one or more of the following methods: (a) spraying the liquid in the air, (b) bubbling air through the liquid, or (c) agitating the liquid to promote surface absorption of air. (2) The supplying of air to confined spaces under napes, downstream from gates in conduits, etc. to relieve low pressures and to replenish air entrained and removed from such confined spaces by flowing water. (3) Relief of the effects of cavitation by admitting air to the section affected.

Aeration porosity. Obsolete term defined as the pore space filled with air when the soil is placed on a porous place and equilibrated with a 50-centimeter hanging water column.

Aerobic. Using or consuming oxygen; the presence of oxygen; opposite of anaerobic.

Aerobic bacteria. Bacteria that require free elemental oxygen for their growth.

Aerobic biodegradation. Biological treatment where microorganisms metabolize biodegradable organics in aqueous waste in an oxygen environment. Includes activated sludge process.

Aerobic digestion. Digestion of suspended organic matter by means of aeration.

Aerosol. System of colloidal particles dispersed in a gas, smoke, or fog.

Afterburner. An off-gas post-treatment unit for control of organic compounds by thermal oxidation. A typical afterburner is a refractory-lined shell providing enough residence time at a sufficiently high temperature to destroy organic compounds in the off-gas stream.

Aggradation. Geological process of building up a surface by deposition.

Aggrading river. A river which is building up its valley bottom by the deposition of material.

Aggregate. Assortment of gravel, pebbles, and/or cinder commonly used as a filler for paving asphalt and concrete.

Aggregation. Formation of aggregates. In drilling fluids, aggregation results in the stacking of the clay platelet face to face; as a result, viscosity and gel strength decrease.

Agitation. The act of exerting force on a solid or liquid mixture for the purpose of inducing mixing and uniform dispersion of the components.

Agitator. Mechanical apparatus for mixing and/or aerating, a device for creating turbulence.

Agric horizon. Subsurface soil horizon which is formed directly under the plow layer and contains silt, clay, and humus.

Air blanks. Air blanks are used for ambient air monitoring and for soil gas sampling. Air blanks may consist of ambient air or hydrocarbon-free air. Air blanks should be run before each soil gas sample to ensure that the sampling train and instruments are clean.

Air-displacement pump. Displacement pump in which compressed air is used as the displacement agency in the cylinders.

Air-filled porosity (f_a). Measure of the relative air content of a soil.

Air filter. A filter of cloth or other material used for removing particulates from air.

Air lift. Device for raising liquid by injecting air in and near the bottom of a riser pipe submerged in the liquid to be raised.

Air-lift pump. A pump that consists of two pipes, one inside the other, and is used to withdraw water from a well. The lower ends of the pipes are submerged, and compressed air is delivered through the inner pipe to form a mixture of air and water. The bubbles mix with the water and reduce the apparent specific gravity of the air/water mixture. The mixture of air and water rises in the outer pipe to the surface, because the specific gravity of the mixture is less than that of the water column.

Air permeability. A coefficient that describes the convective transmission of air through soil in response to a total pressure gradient.

Air saturation. Ratio of the air-filled porosity to total soil porosity.

Air sparging. Groundwater remediation process whereby air is injected into the saturated zone and simultaneously withdrawn from a vapor extraction well in the unsaturated zone. The transport of air bubbles through the groundwater volatilizes contaminants and supplies oxygen to accelerate bioremediation.

Air stripping. Mass transfer process in which a substance in solution in water is transferred to solution in a gas, usually air. It is a type of physical treatment that is used to remove volatile organic contaminants from a liquid to a gas phase.

Albic horizon. Subsurface soil horizon that is light in color.

Albumin. Mixture of water-soluble globular proteins occurring in the tissues and fluids of the body.

Alcohol. Class of organic compounds that contain one or more hydrocarbon groups and one or more hydroxyl groups. Alcohols can be straight or branch-chained (aliphatic) or possess a ring-like structure (i.e., alicyclic, aromatic, heterocyclic, and polycyclic).

Alcoholysis. Chemical reaction between an alcohol and another organic compound.

Aldehyde. An alcohol that is formed by the oxidation of primary alcohols. All aldehydes contain a carbonyl group.

Aldose. Group of sugars whose molecules contain an aldehyde group and one or more alcohol groups.

Aldrin. An agricultural insecticide.

Algae. Chiefly aquatic, eucaryotic, one-celled or multicellular plants without true stems, roots, and leaves which are typically autotrophic, photosynthetic, and contain chlorophyll. Algae are not typically found in groundwater.

Algicide. Substance or chemical used to destroy algal growth.

Alicyclic. Series of cyclic organic compounds, the molecules of which are arranged in a closed ring, and whose properties are more similar to those of an aliphatic than those of an aromatic compound.

Aliphatic. Pertaining to an open-chain carbon compound. Aliphatic compounds are usually associated with petroleum products derived from a paraffin base and having a straight- or branched-chain, saturated or unsaturated molecular structure. Substances such as methane and ethane are typical aliphatic hydrocarbons.

Aliphatic compound. Organic compounds in which the characteristic groups are linked to a straight or branched carbon chain. The name for one of two broad classes of hydrocarbons, aromatic compound being the other, and differentiated from aromatic compounds, which have benzene-type rings. Aliphatic compounds include hydrocarbons, aldehydes, ketones, alcohols, organic acids, and carbohydrates and are either saturated hydrocarbons (only single bonds between carbons) or unsaturated hydrocarbons (double or single bonds between carbons). Examples include propane, ethylene, acetylene, and cyclohexane.

Aliquot. Dividing a sample into two or more equal parts; implies an exact division of a quantity. An aliquot of a field sample (soil or water) is often used for laboratory analysis.

Alkali. Soluble salts, principally of sodium, potassium, magnesium, and calcium, having the property of combining with acids to form neutral salts.

Alkali metal. The univalent metals, which include lithium, sodium, potassium, rubidium, cesium, and francium. These metals are basic in nature and evolve heat when in contact with water.

Alkaline soil. Soil having a pH greater than 7.0.

Alkalinity. Measure of the ability of a liquid to neutralize acids. The alkalinity value, usually reported as milligrams per liter of calcium carbonate, is given for a specific volume. The higher the alkalinity of a body of water, the less likely it is to be affected by acid rain.

Alkalization (solonization). Accumulation of sodium ion on the exchange sites of a soil.

Alkaloid. A group of complex heterocyclic compounds that contain nitrogen. Alkaloids are derived from plants and have strong physiological activity. Alkaloids include nicotine, opium, strychnine, and caffeine.

Alkane hydrocarbons. Organic compounds that contain only carbon and hydrogen.

Alkanes. Hydrocarbon compounds (e.g., $CH_3CH_2CH_3$) that do not contain double or triple bonds between carbons. Alkanes can form straight chains or cyclic structures such as cyclohexane.

Alkenes. Straight-chain hydrocarbon compounds that contain one or more double or triple bonds between carbons.

Alkyd resin. A group of polyester resins used in exterior paints, baked enamels, and marine protective coatings.

Alkyl. A univalent group derived from paraffinic hydrocarbons that contain one less hydrogen than the corresponding hydrocarbons.

Alhynes. The group of unsaturated hydrocarbons with a triple carbon-carbon bond having the general formula C_nH_{2n-2}.

Alloy. Manmade solution of metals, either solid or liquid, which may or may not include a metal.

Alluvial. Deposition of material by streams.

Alluvial clay. Stream deposits of clayey material transported by flowing water.

Alluvial deposit. Sediment deposited in place by the action of streams.

Alluvial fan. The outspread sloping deposit of detrital material brought down by the action of water from neighboring elevations to a plain or open valley bottom.

Alluvial plain. A plain formed by alluvial material eroded from an area of higher elevation.

Alluvial river. A river that has formed its channel by the process of aggradation.

Alluvial soil. Soil formed by the transportation and deposition by streams of material carried a considerable distance before being deposited.

Alluvial terrace. Terraces, usually adjacent to river valleys, which were originally deposited by stream action and later cut through by the stream.

Alluviation. The process of accumulating deposits of gravel, sand, silt, or clay at places in rivers, lakes, or estuaries where the velocity of flow is reduced.

Alluvium. General term for clay, silt, sand, gravel, or similar unconsolidated material deposited during comparatively recent geologic time by a stream or other body of flowing water as a sorted or semi-sorted sediment in the bed of the stream, its floodplain, delta, or as a cone or fan at the base of a mountain.

Allyl alcohol ($CH_2=CHCH_2OH$). A primary alcohol that is created by hydrolyzing allyl chloride with aqueous sodium hydroxide or by the catalytic isomerization of propylene oxide.

Alpha (α). In statistics, a designation of the significance level or probability of a Type 1 error.

Alpha particle. Small, electrically charged particle of very high velocity released by many radioactive materials. It is made up of two neutrons and two protons and has a positive electrical charge.

Alpha radiation. Fast-moving particles that are emitted from the nuclei of radioactive elements.

Alternative hypothesis. Statistical hypothesis that specifies that the underlying distribution differs from the null hypothesis.

Alumel. A nickel-based alloy containing 2.5% magnesium, 2% aluminum, and 1% silicone.

Aluminosilicates. Minerals for which the crystal structures include a tetrahedron of oxygen atoms surrounding a central cation and an octahedron of oxygen or hydroxyl groups surrounding a larger cation.

Aluminum chloride ($AlCl_3$). Inorganic compound that is usually in crystal form. It is formed by reacting chlorine with molten aluminum. It is used as a catalyst in reactions involved in petroleum refining such as cracking, isomerization, and polymerization of aliphatic compounds. During World War II, it was used as a catalyst in the manufacture of synthetic rubbers.

Aluminum sulfate (alum). Water-soluble powder obtained by digesting clay or bauxite with sulfuric acid. It is used in paper coatings, as a flocculating agent in the purification of water and sewage, and in textile dyeing.

Alundum. Porous alumina oxide resembling corundum in hardness. It is manufactured by fusing alumina in an electric furnace and is used chiefly as an abrasive and as a refractive.

Amalgam. Liquid or solid alloy or mixture of mercury with another component that is usually a metal but may be a nonmetal (the component must be soluble in mercury).

Ambient. Prevailing condition in the vicinity, usually relating to some physical measurement such as temperature. Sometimes used as a synonym for background.

Amides. Organic or inorganic compounds that contain nitrogen and are derived from organic acids and ammonia under special conditions.

Amines. An aliphatic compound that is an alkyl derivative of ammonia.

Aminophenol. An aromatic compound derived from nitrophenol or nitrobenzene and used as an organic dye and as a developing agent in photography.

Ammonia. A chemical combination of hydrogen and nitrogen occurring extensively in nature. The combination used in water and wastewater engineering is expressed as NH_3. Ammonia is an uncharged inorganic nitrogen compound with formula NH_3; in water solution, it prevails above pH 9.3. Ammonium is the ion form with formula NH_4^+, which prevails below pH 9.3. Ammonium is attracted to soils with negative charges by the process of ion exchange, thus it tends to move more slowly in groundwater systems than does ammonia.

Ammonia nitrate (NH_4NO_3). Inorganic compound that is made by reacting ammonia with nitric acid. It is used primarily as a fertilizer, although it is also found as an explosive and as an oxidizer in solid rocket fuels.

Ammoniator. Apparatus used for applying ammonia or ammonium compounds to water.

Ammonification. Bacterial process by which organic nitrogen is converted to ammonium.

Amphoteric. Having the characteristic of reacting as either an acid or a base.

Amyl (C_5H_{11}). A hydrocarbon group containing 5 carbon and 11 hydrogen atoms.

Amyl acetate ($CH_3COOC_5H_{11}$). Mixture of several isomers that are derived from either amyl alcohol or oil by reaction with acetic acid. It is used as a solvent in leather polishes, in textile finishing operations, and as a food additive.

Anaerobic. Absence of oxygen.

Anaerobic bacteria. Bacteria that grow only in the absence of free elemental oxygen.

Anaerobic biodegradation. Biological treatment where microorganisms metabolize biodegradable organics in aqueous waste in an oxygen-deficient environment.

Anaerobic digestion. Degradation of organic matter through the action of microorganisms in the absence of elemental oxygen.

Analog. In chemistry, a structural derivative of a parent compound.

Analysis. Testing of substances to determine their chemical composition or hazardous characteristics.

Analysis of variance. Statistical procedure comparing the means of different groups of observations to determine whether there are any significant differences among the groups.

Analyte. Substance in a test sample whose quantity is to be determined or presence detected.

Anemometer. Device that measures the force or velocity of wind. Common types include the rotating vane, the swinging vane, and the hotwire anemometer.

Angle of repose. The greatest angle to the horizontal assumed by any unsupported granular semisolid or semi-fluid material.

Angstrom. Unit of measure often used to measure wavelength of radiant energy. An angstrom is equal to 10^{-8} centimeters or 1 nanometer (nm).

Angular unconformity. A time break in a sequence of depositions where the strata below the unconformity is cut off and is overlain at an angle by the beds after the unconformity.

Anhydride. Compound formed by removing two atoms of hydrogen and one atom of oxygen from another compound.

Anhydrous. Substance in which no water is present in the form of a hydrate or water of crystallization.

Anion. A negatively charged ion in an electrolyte solution.

Anion exchange. The process in which anions in solution are exchanged for other anions from an ion exchanger.

Anionic surfactant. An ionic type of surface-active substance used in cleaning products.

Anisotropic. Characteristic of a material with differing or unequal physical or chemical properties when measured at any given point along its different axes.

Annulus. Space between the drill string or casing and the wall of the borehole or outer casing.

Anomaly. Deviation from a norm for which an explanation is not apparent on the basis of available data.

Anoxic. Total deprivation of oxygen.

Antagonism. (1) Interaction of two substances (e.g., chemicals, pesticides, drugs, or hormones) acting in the same system in such a way that one partially or completely inhibits the effects of the other. (2) Interaction of two types of organism existing in close association in such a way that the growth of one is inhibited by the other.

Antecedent precipitation. Rainfall that occurred prior to the particular rainstorm under consideration.

Antecedent stream. A stream that developed and maintained its course during and after a period of a disturbance.

Anthracite. The hardest and most energy-rich of naturally occurring coals.

Anthropic epipedon. Soil surface horizon that contains over 250 parts per million of citric acid-soluble phosphorus.

Antioxidant. Organic compound that is able to reduce the normal tendency of oxygen to combine chemically with hydrocarbons in petroleum products, animal fats, vegetables oils, and rubber. Antioxidants have the effect of degrading the material to which they are added.

Apparent cohesion. Cohesion of moist soils due to surface tension in capillary interstices.

Apparent color. Color in a solution that is due to suspended matter in the liquid.

Apparent dip. Inclination of a plane measured in a direction not perpendicular to strike.

Apparent dip direction. The bearing of the vertical plane containing the apparent dip angle.

Apparent liquid saturation. The sum of water and oil saturations and of trapped air. It is assumed to be a function of the capillary pressure between air and oil.

Apparent water saturation. The sum of the water and oil trapped in the water phase. It is assumed to be a function of the capillary pressure between oil and water.

Applied action level (AAL). The chemical-specific, medium-specific concentration established by the California Department of Health Services (DHS) which is not expected to cause adverse effects in humans exposed to this concentration 24 hours/day, 7 days/week, for a period of 70 years.

Aquatic species (organisms). Organisms, plants, and animals living in water or whose habitat needs (spawning, nesting, feeding resting) include the water medium.

Aqueous solubility. Extent to which a compound will dissolve in water. The log of solubility is generally inversely related to molecular weight.

Aquic moisture regime. Soil sufficiently saturated so that reducing conditions exist.

Aquiclude. A saturated but poorly permeable bed, formation, or group of formations that does not yield water freely to a well or spring; may transmit appreciable water to or from adjacent aquifers.

Aquifer. Formation, group of formations, or part of a formation that contains sufficient saturated permeable material to yield significant quantities of water under ordinary hydraulic gradients.

Aquifer test. Test involving the withdrawal of measured quantities of water from or addition of water to a well, and the measurements of resulting changes in head in the aquifer both during and after the period of discharge or addition.

Aquifuge. Rock without interconnected openings which can neither absorb nor transmit water.

Aquitard. Geologic formation, or group of formations, or part of a formation through which virtually no water moves.

ARAR. Acronym for Applicable Relevant and Appropriate Requirements. ARARs include the federal standards and more stringent state standards that are legal or relevant and appropriate under the circumstances. ARARs include cleanup standards, standards of control, and other environmental protection requirements, criteria, or limitations. RCRA has frequently been used as an ARAR for cleanup of Superfund sites.

Area of influence. Area within which the potentiometric surface is lowered by withdrawal or raised by injection of water through a well.

Arene. Class name for unsaturated cyclic (aromatic) compounds. Benzene is an example.

Argillic horizon. Subsurface soil horizon that has at least 1.2 times more clay than the overlying horizon; it is formed by the illuviation of clay from overlying soils.

Arid. (1) Term applied to regions where precipitation is so deficient in quantity or occurs at such a time that agriculture is impracticable without irrigation; (2) climate that has insufficient rainfall to support vegetation.

Aridic moisture regime. Soils that are dry more than half of the time when not frozen. They are never moist more than 90 consecutive days when soil temperatures are above 8°C at 50 centimeters in depth.

Arithmetic average. Sum of a set of observations divided by the number of observations.

Aromatic. (1) Fragrant; spicy; strong-scented; odoriferous; having an agreeable odor. (2) A series of benzene ring compounds, many of which have an odor or are derived from materials having an odor.

Aromatic compounds. Organic compounds that have ring structures or cyclic groups in their structure.

Aromatic hydrocarbons. Compounds containing one or more benzene rings (a six-carbon ring structure with alternating double bonds between carbons).

Arrhenius equation. An equation used to extrapolate the reaction rate of a high-temperature reaction to a lower rate reaction. An expression of this equation is

$$\ln k_2/k_1 = E_a/R \ (T_2 - T_1/T_1T_2)$$

where

k = a rate constant.
E_a = the activation energy (Kj/mol).
R = universal gas constant (8.314 J/mol^{-K}).
T = temperature in degrees Kelvin. **Arroyo**. A stream channel or gully, usually rather small, walled with steep banks, and dry much of the time.

Arsenic (As). A highly poisonous metallic element. Arsenic and its compounds are used in insecticides, weed killers, and industrial processes. Arsenic occurs in two environmentally significant valence states, As^{+3} or As^{III} (trivalent) and As^{+5} or As^V (pentavalent), with different toxic properties. The various organic forms of arsenic include methylated forms, arseno-lipids, arseno-sugars, arseno-bentaine, and arseno-choline.

Arsenic trioxide (As_2O_3). A white powder that is extremely toxic and is used as a decolorizing agent in the manufacture of glass, as a pesticide, and as a preservative of leather in the tanning industry.

Artesian aquifer. Aquifer that is overlain by a confining layer and is under pressure.

Artesian discharge. Rate of discharge of water from a flowing well.

Artesian flow. Flow from an underground-water-bearing stratum under sufficient pressure to force water above ground level.

Artesian head. The distance above or below the land surface to which the water in an artesian aquifer or groundwater basin would rise if free to do so.

Artesian pressure surface. A piezometric surface having a position above the upper surface of the saturation zone.

Artesian spring. A spring in which water issues under pressure through some fissure or other opening in the confining formation above the aquifer.

Artesian water. Subsurface water under sufficient pressure to cause it to rise above the bottom of the superimposed confining formation.

Artesian well. A well tapping a confined or artesian aquifer in which the static water level is above the ground surface.

Artesian well capacity. Rate at which a well will yield water at the surface as a result of artesian pressure.

Artificial recharge. Replenishment of the groundwater supply by means of spreading basins, recharge wells, irrigation, or induced infiltration.

Asbestos. Fibrous magnesium silicate; highly carcinogenic.

Asbestos-cement pipe. Pipe made of a mixture of asbestos fiber and Portland cement combined under pressure.

Askarel®. Generic name for a broad class of nonflammable chlorinated hydrocarbons that are used as dielectric fluids in transformers, capacitors, and special power cables. Askarel fluids contain polychlorinated biphenyls (PCBs).

Asphalt. Thick viscous mixture of hydrocarbons found in the residue from petroleum distillation. Asphalt contains hydrocarbons and a small percentage of heterocyclic compounds that contain sulfur and nitrogen.

Asphalt incorporation. Soil containing volatile organic compounds that are incorporated into hot asphalt mixes as a partial substitute for stone aggregate.

Assay. Analysis of the proportions of metal in an ore. The composition of the ore and its purity, weight, or other properties of economic interest are tested.

Assessment. The quantitative and qualitative study of hazardous waste and its sources; may include its effects on the environment, including the human population.

ASTM. American Society for Testing and Materials, a technical organization which develops standards on characteristics and performance of materials, products, systems, and services. It is the world's largest source of voluntary consensus standards.

Asymptote. A line that is considered to be the limit to a curve. As the curve approaches the asymptote, the distance separating the curve and the asymptote continues to decrease, but the curve never actually intersects the asymptote.

Atomic number. Number of protons in the nucleus of an atom on which its structure and properties depend. The atomic number signifies the location of the element in the Periodic Table of the Elements and is the same as the number of negatively charged electrons

Atomic weight. Reference to the relative weights of atoms as compared to some standard.

Atmometer. Instrument for measuring evaporation.

Atrazine. A selective herbicide used for season-long weed control in corn, sorghum, and certain other crops; also used for nonselective weed control in areas without crops. Its chemical name is 2-chloro-4-ethylamino-6-isopropylami-no-5-triazine.

Attenuation. General term that relates to the reduction in magnitude, intensity, or concentration of a substance.

Attenuation vs. distance relationship. Curve showing the decay of the amplitude of the seismic wave with distance from the earthquake source.

Atterberg limits. The water content of a soil as it passes from a solid state when dry, through the semisolid, plastic, and liquid states as water is added. Atterberg limits are used in soil classification and form the basis for the soil groups defined by the Unified Soil Classification System.

Auger. A tool for drilling/boring into unconsolidated earth materials (soil) consisting of a spiral blade wound around a central stem or shaft that is commonly hollow (hollow-stem auger). Augers commonly are available in flights (sections) that are connected together to advance the depth of the borehole.

Autoignition temperature. Temperature at which a substance will spontaneously ignite. Autoignition temperature is an indicator of thermal stability for petroleum hydrocarbons.

Autotroph. Ability of an organism to create living matter from inorganic raw material.

Autotrophic. Bacteria that can oxidize inorganic material.

Autotrophic organism. An organism that does not require organic food as a substrate.

Autoxidation. Oxidation that occurs spontaneously in a substance that is exposed to air or oxygen at temperature below 300°F.

Available moisture. Water easily abstracted by root action and limited by field capacity and the wilting coefficient.

Available oxygen. The quantity of dissolved oxygen available for oxidation of organic matter.

Average. Arithmetic mean obtained by adding quantities and dividing the sum by the number of quantities.

Average stream flow. The average rate of discharge during a specified period.

Avogadro constant. Physical constant equal to 6.0225×10^{23} mol^{-1}. Amadeo Avogadro (1776–1856) advanced the principle that the number of molecules of any gas at 32°F for a given volume is the same regardless of the chemical composition or physical properties of the gas.

Azide. Group of explosive compounds whose molecules consist of a chain structure of double-bonded nitrogen atoms that are attached to a metal or metal complex, halogen, hydrogen, ammonium radical, or organic radical. Lead azide — $Pb(N_3)_2$ — is an example.

B

Backflow preventer. Device for a water supply pipe to prevent the backflow of water into the water supply system.

Background level. Normal ambient environmental concentration of a substance.

Backwash. Reversal (downward) flow of water in well to remove fines and enhance well production.

Backwash (filter wash). Reversal of flow through a rapid sand filter to wash clogging material out of the filtering medium and reduce conditions causing loss of head.

Bacteria. Organisms that do not produce chlorophyll.

Baffle aerator. Aerator in which baffles are provided to cause turbulence and minimize short-circuiting.

Bag filter. A closely woven fabric bag used to remove particulates from a gas stream that passes through the bag.

Bagasse. The cellulosic residue of sugar cane after it has been processed to remove the sugar.

Baghouse. A structure containing bag filters that are used to remove particulates from a gas stream.

Bailer. Open-ended tubular hollow device with a valve at or near the bottom end which can be closed and is designed to retrieve water samples from a monitoring well.

Ball valve. Valve regulated by the position of a free-floating ball that moves in response to fluid or mechanical pressure.

Banded steel pipe. A steel pipe, the strength of which has been increased by the use of bands shrunk around the shell.

Bar. (1) Unit of pressure equal to one million dynes per square centimeter or 0.98 atmospheres; (2) alluvial deposit or bank of sand, gravel, or other material at the mouth of a stream or at any point in the stream itself which causes an obstruction to flow or to navigation.

Bar screen. Sequence of regularly spaced metal bars used to remove solids from sewage effluent.

Barite. The natural form of barium sulfate found in the southwestern United States. Its primary use is to increase the weight of muds used to drill oil wells.

Barium sulfite (barite). A heavy white powder made synthetically by reacting barium chloride with sodium sulfate. It is used as a weighing agent in drilling muds and as a weighing filler in low-grade electric tapes and plastic products.

Barn. Unit of measurement equal to 10^{-24} square centimeters. It is used by physicists to describe the target cross-section area of an atomic nuclei.

Barometer. Instrument used to measure air pressure.

Barrel. Volumetric unit of measurement used in the petroleum industry. One barrel is equal to 42 U.S. gallons, 35 Imperial gallons, or 159 liters.

Barrier spring. A spring which occurs when a raising of the confining bed forces groundwater to rise to the ground surface.

Barrier well. A well installed to intercept and pump out a plume of contaminated groundwater.

Basalt. Fine-grained igneous rock usually occurring in volcanic flows, dikes, and sills.

Base. Material that produces hydroxide ions when dissolved in water.

Base exchange. An ion exchange process whereby calcium and magnesium ions are exchanged for sodium ions.

Base flow. That part of the stream discharge that is not attributable to direct runoff from precipitation or melting snow.

Base line. A surveyed line established with unusual care to which surveys are referred for coordination and correlation.

Basin. Natural or artificially created space or structure, surface or underground, which has a shape and character of confining material that enables it to hold water.

Basin plan. A plan for the protection of water quality prepared by the California Regional Water Quality Control Board in response to the Porter Cologne Water Quality Act.

Batch. A group of samples analyzed sequentially using the same calibration curve, reagents, and instrument.

Batch process. An industrial procedure in which raw materials are mixed in individual groups or batches.

Bating. The process used in the manufacture of leather in which dehaired skins or hides are treated with enzyme-containing materials to soften the leather. Enzymes used include pancreatin and trypsin.

Bauxite ($Al_2O_3(3H_2O)$). A mineral whose primary component is hydrated aluminum oxide. Bauxite is the chief ore of aluminum.

Bearing. Horizontal angle between a line and a specified coordinate direction, usually north or south.

Bearing capacity. The load per unit area which a soil can safely support without excessive yield.

Bed. Bottom of a watercourse or any body of water.

Bed load. Sediment that moves by sliding, rolling, or skipping on or very near the stream bed.

Bedding. The arrangement of a sedimentary rock in beds or layers of varying thickness and character; the general physical and structural character or pattern of beds and their contacts with the rock mass. The term may be applied to the layered arrangement and structure of an igneous or metamorphic rock.

Bedding plane. The plane or surface separating individual layers, beds, or strata in sedimentary or stratified rocks.

Bedload. The part of the total stream load that is moved on or immediately above the stream bed; that part of the stream load that is not continuously in suspension or solution.

Bedrock. General term for the rock, usually solid, that underlies soil or other unconsolidated material.

Beer's Law. A law that deals with the absorption of light as it passes through a solution. The law states that the intensity of monochromatic light passing through a solution exponentially decreases as the concentration of the absorbing medium increases.

Bell-and-spigot joint. Form of joint used on pipes which have an enlarged diameter or bell at one end and a spigot at the other which fits into and is laid in the bell.

Belt filter press. Filtration process in which a press continuously squeezes sludge through a series of rollers which apply increasing pressure and shear force on the sludge.

Belt screen. Endless band or belt of wire mesh, bars, plates, or other screening medium which passes around upper and lower rollers as guides.

Beneficial uses. "Beneficial uses" of the waters of the state that may be protected against quality degradation include, but are not limited to, domestic, municipal, agricultural, and industrial supply; power generation; recreation; esthetic enjoyment; navigation; and preservation and enhancement of fish, wildlife, and other aquatic resources or preserves. Equivalent to "designated uses" under federal law.

Bentonite. A colloidal clay that is largely composed of the mineral montomorillonite and is an absorptive clay derived from decomposed volcanic ash. It is often used as a grout material (sealant), is able to absorb large quantities of water, and can expand to several times its normal volume.

Benzene (C_6H_6). An aromatic compound composed of six carbon atoms arranged in a hexagonal ring. A hydrocarbon atom is attached to each of the carbon atoms. It is manufactured from toluene by hydrodealkylation or from coal tar by fractional distillation. Benzene is used to produce a variety of aromatic compounds used as plastics, detergents, elastomers, insecticides, solvents, and dyes. Chemicals manufactured from benzene include nitrobenzene, ethylbenzene, dichlorobenzene, benzene sulfonic acid, styrene, phenol aniline, and cyclohexane.

Benzenesulfonic acid ($C_6H_5SO_3H$). An aromatic acid made by reacting concentrated sulfuric acid with benzene. It is used in the manufacture of organic chemicals such as phenol, resorcinol, and dyes.

Benzo-α-anthracene. Specific isomers of the benzoanthracenes which are polynuclear aromatic hydrocarbons and are listed by the EPA as a priority pollutant under Section 307(a) of the Clean Water Act.

Benzo-α-pyrene. A polynuclear aromatic hydrocarbon which is one of the 126 priority pollutants listed by the EPA under Section 307(a) of the Clean Water Act.

Benzoic acid (C_6H_5COOH). An organic acid created by attaching a carboxyl group to a benzene ring. It is made from toluene either by oxidation or by chlorination to benzotrichloride, followed by hydrolysis. It is used as a vulcanization retarder for rubber and as an intermediate in organic synthesis.

Benzopyrene (C_2OH_{12}). The class of polynuclear aromatic hydrocarbons which contain five joined benzene rings.

Benzyl alcohol ($C_6H_5CH_2OH$). An aromatic alcohol obtained by hydrolyzing benzyl compounds such as chloride or acetate. It is used as a solvent for cellulose esters, waxes, and resins and is found in soaps, inks, and some coating formulations.

Berm. A horizontal strip or shelf built into an embankment or cut to break the continuity of an otherwise long slope.

Bernoulli equation. An equation used to describe energy loss during fluid flow. One expression for this formula is

$$H_T = \rho/\gamma + V^2/2g + z$$

where

H_T = total head or energy loss per unit mass.
ρ = mass density.
γ = weight of water.
V^2 = velocity.
g = acceleration due to gravity.
z = elevation head.

Best demonstrated available technology (BDAT). The technology EPA establishes for a land-banned hazardous waste to reduce overall toxicity or mobility of toxic constituents in the waste. BDAT must be applied to such a waste prior to land disposal unless one can successfully demonstrate the validity of an equivalent treatment method.

Best management practices (BMPs). A practice or combination of practices determined after an examination of alternative practices and appropriate public participation to be the most effective and practical (including technological, economic, and institutional considerations) means of preventing or reducing the amount of pollution generated by nonpoint sources to a level compatible with water-quality goals.

Beta particle. A small, electrically charged particle emitted by the atomic nucleus of many radioactive materials during radioactive decay; identical to the electron. Beta particles emerge from the radioactive material at high speeds.

Beta radiation. Electrons that consist of negatively charged particles moving at speeds ranging from 30 to 99% of the speed of light. Electron velocity is a function of the given element emitting the electron.

Bicarbonate alkalinity. Alkalinity caused by bicarbonate ions.

Binding energy. The energy that holds the protons together in an atomic nucleus.

Bioaccumulative. Characteristic of a chemical species when the rate of intake into a living organism is greater than the rate of excretion or metabolism. This results in an increase in tissue concentration relative to the exposure concentration.

Bioassay. The employment of living organisms to determine the biological effects of a substance, factor, or condition.

Bioassessment. An assessment of the condition of a water body using any available biological methods. Biosurvey and bioassay are common bioassessment methods.

Bioaugmentation. The introduction of cultured microorganisms into the subsurface environment for the purpose of enhancing bioremediation of organic contaminants. Generally, the microorganisms are selected for their ability to degrade the organic compounds present at the remediation site. The culture can be either an isolated genus or a mix of more than one genera. Nutrients are usually also blended with the aqueous solution containing the microbes to serve as a carrier and dispersant. The liquid is introduced into the subsurface under natural conditions (gravity fed) or injected under pressure.

Bioavailability. Availability of a compound for biodegradation, influenced by the location of the compound relative to microorganisms and its ability to dissolve in water.

Bioavailability potential. Measure of the relative bioavailability of different elements, ions, radicals, or molecules.

Biochemical oxygen demand (BOD). Biochemical test used in the water-treatment industry to test for the presence of organic matter in a water sample that can be degraded in the presence of oxygen.

Biochemical process. The process by which the life activities of bacteria and other microorganisms degrade complex organic material into simple, more stable substances.

Biochemistry. The science of chemical changes brought about by living organisms.

Biocide. Substance capable of destroying (killing) living organisms.

Bioconcentration. Positive difference in concentration of a chemical between water and that in an organism living in that body of water due to direct uptake of the chemical from the water.

Bioconcentration factor (BCF). The ratio between the chemical contaminant concentration in aquatic animal tissue and that in water. Established BCFs are found in the technical literature; values are summarized in the Superfund Public Health Evaluation Manual of the EPA.

Biodegradability (biodegradation potential). The relative ease with which petroleum hydrocarbons will degrade as the result of biological metabolism. Although virtually all petroleum hydrocarbons are biodegradable, biodegradability is highly variable and dependent somewhat on the type of hydrocarbon. In general, biodegradability increases with increasing solubility; solubility is inversely proportional to molecular weight.

Biodegradable. Pertaining to materials that can be decomposed by microorganisms.

Biodegradation. (1) Susceptibility of an organic material to decomposition as a result of microbial attack; (2) destruction or mineralization of either natural or synthetic organic materials by microorganisms populating soils, natural bodies of water, or wastewater treatment systems.

Biogas. Gas formed by the anaerobic decomposition of organic matter. Methane and carbon dioxide are the primary components.

Biological filter. In the sewage treatment industry, a bed of inert granular material over which sewage is introduced and allowed to percolate. The aerobic bacteria and fungi on the granular material reduce the biochemical oxygen demand of the effluent.

Biological filtration. The process of passing a liquid through the medium of a biological filter, thus permitting contact with attached bacterial films that adsorb and adsorb fine suspended, colloidal, and dissolved solids and release end products or biochemical action.

Biological transformation. Structural alteration of a chemical by an organism.

Biological treatment. A technology process whereby the organic components of hazardous waste are biologically decomposed or altered under controlled conditions to a state which is either nontoxic or less toxic than its pretreatment state. Composting is an example.

Biomagnification. The net accumulation and increase of a substance in an organism as a result of consuming organisms from lower trophic levels (e.g., the consumption of algae by fish or water plants by ducks).

Biomass. The amount of living matter in a given area or volume.

Biophotolysis. Electrons produced in the first stages of photosynthesis which can be used to create free hydrogen.

Bioremediation. A hazardous waste site remediation technique that utilizes micro-organisms to metabolize hazardous organic constituents in waste to nonhazardous compounds. Environmental conditions are carefully controlled in an attempt to create optimum growth conditions for the organisms. Bioremediation can be often implemented *in situ*.

Biosphere. Term used to include all living matter on the planet.

Biota. All living organisms that exist in an area.

bis **(2-ethylhexyl) phthalate.** Also known as di-sec-octyl phthalate; a light-colored, viscous, odorless, combustible liquid.

Bismuth. A brittle, grayish-white, red-tinged metallic element used in the manufacture of fusible alloys.

Bismuthinite. A mineral, bismuth sulfide (Bi_2S_3), occurring in lead-gray masses; an ore of bismuth.

Bitumen. A semisolid or solid form of petroleum.

Bituminous. Form of coal that is softer than anthracite but harder than lignite. Bituminous coal usually contains a high proportion of volatile compounds (13%).

Blank result. Result of the analysis of a method blank, which is reagent water analyzed using the same reagents, instruments, and procedures as the samples in a batch. A method blank is used to detect laboratory contamination.

Blanks. Blanks are run as a quality control measure to show that any contamination detected in a sample is associated with the sample and has not been introduced by the operator (via a dirty syringe or contaminated solvents) or the instrument (via contaminated inlet lines).

Bleach. To remove or decolorize impurities in a material. It usually involves a chemical reaction such as oxidation or reduction and is used in textile, paper pulp, vegetable and mineral oils, and clays.

Blow count. Number of blows of a 140-pound hammer falling 30 inches to drive a 2-inch outer diameter (1.38-inch inner diameter) split spoon sampler. The blow count is qualitatively associated with the relative density and consistency of a soil. The following blow counts per foot are associated with the relative densities of sands and gavels:

Sands and Gravels	Blows/Foot
Very loose	0–4
Loose	4–10
Medium dense	10–30
Dense	30–50
Very dense	over 50

The consistency of silts and clays is associated with the following blows per foot.

Silts and Clays	Blows/Foot
Very soft	0–2
Soft	2–4
Firm	4–8
Stiff	8–16
Very stiff	16–32
Hard	over 32

Blue. Compound which has a pigmenting effect and may be derived from cyanide-containing compounds.

Blue copperas. Copper sulfate.

Blue gas. A gas rich in carbon monoxide and hydrogen (heating value of about 300 Btu per cubic foot) made by alternately combusting a bed of coal with air and reacting the hot coke formed with steam; sometimes called water gas.

Blue-green algae. Algae that contains chlorophyll and red and blue pigments.

Blue stone. Common name for copper sulfate.

BOD (biological/biochemical oxygen demand). Amount of oxygen used by bacteria while decomposing organic matter in the presence of oxygen. BOD is often used as a measure of the pollution strength of wastes based on the oxygen demand of the bacteria to decompose the waste.

Bodied oil. A vegetable oil whose viscosity has been increased by heating.

Boiler feedwater. Water forced into a boiler to take the place of that evaporated in the generation of steam.

Boiler scale. An encrustation deposited on boiler heating surfaces by precipitation of minerals from the water used.

Boilers. Vessels where hazardous waste may be used as supplementary fuel to coal or oil.

Boiling point. Temperature at which a component's vapor pressure equals atmospheric pressure. Boiling point is a relative indicator of volatility and generally increases with increasing molecular weight.

Bolt ring. A closing device used to secure a cover to the body of an open-head drum. This ring requires a bolt and nut for the closure.

Bond number. A dimensionless ratio between gravitational and capillary forces.

Booster pump. A pump installed on a pipeline to raise the water pressure on the discharge side of the pump.

Borehole. An uncased drill hole.

Borings. A borehole is drilled down to a desired depth, material is removed to be examined at the surface, and the elevations at which changes in material are found are recorded in a boring log.

Boron (B). A nonmetallic, chemical element produced in the form either of a brown amorphous powder or very hard, brilliant crystals; its compounds are used in the preparation of boric acid, water softeners, soaps, enamels, glass, and pottery.

Borosilicate. A heat-resistant glass whose principal ingredients are boric oxide, silica, soda, and lime.

Boulder. A rock with a diameter larger than 8 inches.

Bound residue. An organic residue in soil that is unextractable after sequential extractions of the soil with organic solvents.

Bound water. A thin film of water adsorbed on a molecule or other particle as a result of electrical affinity.

Bourdon gauge (Bourdon tube gauge). A semicircular or coiled flexible metal tube attached to a gauge that records the degree to which the tube is straightened by the pressure of the gas or liquid inside.

Boyle's Law. States that the volume of any confined gas at constant temperature varies inversely to the pressure applied to it.

Brackish. Characterized as a salt solution less concentrated than that of seawater (brackish water has approximately 10,000 to 30,000 mg/L of dissolved solids).

Braided river. A river of which the channel is extremely wide and shallow and the flow passes through a number of small, interlaced channels separated by bars or shoals.

Braided stream. A stream that divides into or follows an interlacing network of several small, branching, and reuniting shallow channels. These channels are separated from each other by branch islands or channel bars.

Branched chain. Linear series or chain of carbon atoms occurring in hydrocarbons or alcohols.

Brass. Alloy of copper and zinc, often containing tin. Brass is corrosion resistant.

Brass sleeves. Brass cylinders inserted inside a soil sampling device for collecting a soil sample for physical or chemical testing.

Brazing. A type of welding in which nonferrous alloys are used as filler materials between the pieces of metals to be welded; brazing is performed at temperatures above 1000°F.

Breccia. Rock that has been broken due to the movement of a fault.

Bridging. A condition within the filter pack outside the well screen of a monitoring well whereby the smaller particles are wedged together in a manner that causes blockage. It may result in void spaces below this area in the filter pack.

Brine. Concentrated brackish saline or sea water containing more than 36,000 mg/L of total dissolved solids.

Broad-spectrum pesticide. Pesticide that kills a wide range of pest species.

Brownian movement. The random movement of microscopic particles suspended in a gaseous or liquid suspension.

Btu. The quantity of heat required to raise the temperature of 1 pound of water 1°F at 39°F; used as the standard for the comparison of heating values of fuels.

Bubble radius. The maximum radial distance away from a biosparging well where the effects of sparging are observable. Analogous to radius of influence of an air-sparging well.

Bucket dredge. A dredge which lifts excavated material by an endless chain of buckets or a single drag bucket.

Buffer. A substance or mixture of compounds which, when added to a solution, is capable of neutralizing both acids and bases without appreciably changing the original acidity or alkalinity of the solution.

Buffering capacity. The buffering index β of a solution is defined as the slope of a titration curve of pH vs. moles of strong bases added C_B or moles of strong acid added (C_A):

$$\beta = \frac{dC_B}{dpH} = \frac{-dC_A}{dpH}$$

The buffer index indicates the number of moles of acid or base required to produce a prescribed pH change.

Bulk density (ρ). The density of a material measured in mass per unit volume.

Bulk density, soil. The mass of a known soil volume compared to an equal volume of water, or weight per unit volume.

Bulk modulus (incompressibility modulus). A modulus of elasticity which relates a change in volume to the hydrostatic state of stress. It is the reciprocal of compressibility.

Bulk specific gravity. The ratio of the bulk density of a soil to the mass unit volume of water.

Bulk volume. The volume, including the soils and the pores, of a soil mass.

Buna rubbers. Synthetic rubbers produced from butadiene with a sodium catalyst.

Bung. A threaded closure that is located in the head or body of a drum.

Buoyancy. The upward force that acts on a body when it is totally immersed in a fluid.

Burette. A liquid measuring device used in chemical laboratories; it is a vertical glass tube, open at the top and supported by a bracket. It is equipped with graduated marks on its side and a hand-operated stopcock at or near the bottom.

Buried stream channel. A former stream channel that has been filled with alluvial or glacial deposits and later covered by other material to a degree that prevents its existence or location from being readily discerned from surface indications.

Butane (C_4H_{10}). A saturated hydrocarbon gas that is obtained from petroleum refining. It is used in the production of elastomers and is an organic intermediate.

Butte. An isolated, flat-topped hill or mountain with steep sides common in arid regions.

Butterfly valve. A valve wherein the disk, as it opens or closes, rotates about a spindle supported by the frame of the valve.

Butyl rubber. Synthetic elastomer produced by the catalytic copolymerization of isobutene. As a component of latex, it is used to insulate wires and cables and to coat fabrics and composites.

C

Cable-tool drilling. Method of drilling wells by the use of cable tools. The hole is drilled by a heavy bit, which is alternately raised by a cable and allowed to drop, breaking and crushing the material which it strikes.

Cadmium (Cd). Soft, bluish-white metallic element known to cause cancer in animals, although it is not a confirmed human carcinogen.

Caisson. Watertight chamber consisting of a wood, steel, plastic, or concrete, open at the bottom and containing air under enough pressure to prohibit the entrance of water.

Calcic horizon. A subsurface soil horizon, at least 15 centimeters thick, that has a secondary accumulation of carbonates.

Calcination. The heating of an inorganic material to a temperature below its melting point but high enough to force the evolution of carbon dioxide and bound water along with other impurities by oxidation and reduction. Calcination is used to heat limestone, gypsum, magnesium carbonate, and metallic ores.

Calcium carbonate ($CaCO_3$). Inorganic substance that is the major component of limestone, marble, chalk, and oyster shells and exists in a relatively pure state as calcite.

Calcium oxide (CaO). Grayish-white powder created by heating limestone; also called quicklime or lime.

Calcium sulfate ($CaSO_4$). Grayish-white powder occurring in an anhydrous and hydrated form; hydrated form is known as gypsum.

Calibrate. To check, fix, or correct a measuring instrument to agree with a reference standard.

Caliche. A hard deposit consisting mostly of calcium carbonate or of gravel and sand cemented by calcium carbonate; found in the subsoil in arid sections.

California list of wastes. A RCRA hazardous waste that contains cyanides, RCRA metals, or PCBs or is corrosive (i.e., pH of 2 or less); any liquid or nonliquid hazardous waste containing halogenated organic compounds (HOCs).

Calorie. The quantity of heat which, when added to a unit weight of water, raises the temperature 1°C.

Calorimetry. Measurement of the heat gained or lost by a substance when its physical state is altered.

Cambric horizon. Subsurface soil horizon consisting of very fine sand, loamy fine sand, or finer texture that has less than 1.2 times the clay content as the overlying soil horizon.

Cantena. A sequence of soils of about the same age derived from similar parent materials and climatic conditions but which have different characteristics due to variations in drainage and relief.

Capacitance. The electrical property of a nonconductor that allows the storage of energy as a result of electric displacement when opposite surfaces of the nonconductor are maintained at a difference in potential.

Capacitor. An object consisting of two sheets of equally charged conducting metal of opposite signs and separated by a thin film of dielectric material; also called a condenser. A capacitor is used to measure capacitance, as defined by:

$$C = \frac{Q}{E}$$

where

C = capacitance.
Q = charge in coulombs.
E = voltage across the capacitor.

Capillarity. The phenomenon demonstrated by tubes with minute openings which, when immersed in a fluid, raise or depress the fluid in the tubes above or below the surface of the fluid in which they are immersed.

Capillary fringe. The zone immediately above the water table in which all or some of the interstices are filled with water that is under less than atmospheric pressure and is continuous with the water below the water table. The water is held above the water table by interfacial forces (surface tension).

Capillary head. Difference between capillary lift and the position of the meniscus in a capillary opening.

Capillary interstice. Opening small enough to produce appreciable capillary rise.

Capillary lift. The height to which water or other liquid rises in a capillary tube.

Capillary number. A dimensionless ratio of viscous to capillary forces.

Capillary potential (matric potential). The amount of work that must be done per unit quantity of pure water in order to transport, reversibly and isothermally, an indeterminate quantity of water (identical in composition to the soil water) from a pool and at the external gas pressure of the point under consideration to the soil water.

Capillary pressure. The difference in pressure across the interface between two immiscible fluid phases concurrently occupying the interstices of any porous material. This difference is due to the tension of the interfacial surface, the value of which depends upon the surface curvature.

Capillary suction. The process by which water rises above the water table into the void spaces of a soil due to tension between the water and soil particles.

Capping. The covering of a landfill or injection well, etc. after operations have ceased.

Carbide. A compound formed by heating a carbon and a metal or non-metal material in an electric furnace. Carbide materials are extremely hard and resistant to high temperatures. Examples are calcium carbide, silicon carbide, tungsten carbide, and aluminum carbide.

Carbohydrate. A class of compounds that include simple sugars, starches, and cellulose. Carbohydrate molecules are composed of carbon, hydrogen, and oxygen with the ratio of hydrogen and oxygen the same as in water.

Carbon black. A type of amorphous carbon that is characterized by a small particle size that results in a surface area as high as 18 acres per pound. It is used by the rubber industry in tire treads and as a pigment in paints, printing inks, and plastics.

Carbon monoxide (CO). Colorless gas formed from the incomplete combustion of a substance.

Carbon tetrachloride (CCl_4). Chlorinated hydrocarbon in which the hydrogen atoms of methane are replaced by chlorine. It is a heavy liquid used as a solvent for heavy oils, crude rubber, and greases; as a cleaning agent for metal surfaces and a fumigant; and in the manufacture of refrigerants.

Carbonate. (1) Sediment formed by organic or inorganic precipitation from an aqueous solution of carbonates of calcium, magnesium, or iron; (2) compound formed by the reaction of carbonic acid with an organic compound or metal.

Carbonate hardness. The total hardness of water that is chemically equivalent to the bicarbonate plus carbonate alkalinities of the solution.

Carbonate rocks. A rock consisting chiefly of carbonate minerals, such as limestone and dolomite.

Carboniferous. A geologic period between 360 and 286 million years ago; epochs belonging to this period include the Pennsylvanian and Mississippian.

Carbonization. Treatment of coal or other solid fuel with heat in a closed vessel for the purpose of producing coke or gas.

Carboxyl. A univalent group or radical that is characteristic of an organic acid. It is often shown in formulas as –COOH. The hydrogen atom in the carboxyl can be replaced by other elements in reactions involving the formation of metal salts and soaps.

Carburetion. The process of enriching the heating value of a gas by mixing it with volatile light hydrocarbons, usually produced by thermal cracking of oil.

Carcinogen. A chemical substance or physical agent (such as radiation) that increases the incidence of cancer.

Carcinogenic polycyclic aromatic hydrocarbons (PAHs). These include the following chemicals: benzo(a)pyrene, benzo(a)anthracene, benzo(b)fluoranthene, benzo(k)fluoranthene, benzo(g.h.i)perylene, chrysene, dibenzo(a.h)anthracene, and indeno(1,2,3-cd)pyrene.

Carrier gas. Gas used to transport a gaseous sample through a chromatographic column to the detector of a gas chromatograph. This type of gas is commonly referred to as "ultra zero grade" or "hydrocarbon free" air.

Carrying capacity. Maximum population of plants and/or animals that a given area of land can sustain indefinitely.

Case control study method. A study method in which individuals with a particular disease are compared with similar but nondiseased individuals in regard to their exposure history to a suspected agent. The proportion of the cases with a history of exposure to the agent is compared with the proportion of nondiseased individuals.

Casing. Steel or plastic tubing welded or threaded together to line a borehole.

Castone. Dental powder.

Catalysis. A substance that can retard or accelerate the rate of a chemical reaction.

Catalyst. A substance that has the ability to increase the rate of a chemical reaction.

Catalytic oxidizer. An off-gas post-treatment unit for control of organic compounds. Gas enters the unit and passes over a support material coated with a catalyst (commonly a noble metal such as platinum or rhodium) that promotes oxidation of the organics. Catalytic oxidizers can also be very effective in controlling odors. High moisture content and the present of chlorine or sulfur compounds can adversely affect the performance of the catalytic oxidizer.

Catalytic reactor. Equipment attached to the exhaust pipes of an internal combustion engine which converts uncombusted hydrocarbons and carbon monoxide to carbon dioxide and water vapor.

Cathodic protection. A technique to prevent the corrosion of a metal surface by making that surface the cathode of an electrochemical cell.

Cation. An ion with a positive charge that migrates toward a negative cathode in a solution when an electric current is applied.

Cation exchange. An ion exchange process in which cations in solution are exchanged for other cations from an ion exchanger.

Cation exchange capacity (CEC). The capacity of a soil to sorb cations and to exchange species of these ions in reversible chemical reactions.

Caustic alkalinity. Alkalinity caused by hydroxyl ions.

Cell. (1) Device used to produce an electric current by chemical means or for performing electrolysis; (2) basic unit of tissue structure; (3) hollow structural unit of a plastic foam.

Cellulose. A carbohydrate polymer formed through photosynthesis.

Cellulose acetate. A reaction product of acetic acid and of cellulose. It is used in solutions of photographic film, magnetic tape, and textile products.

Cement, Portland cement. A cement that contains oxides of calcium, aluminum, iron, and silicon made by heating a mixture of limestone and clay in a kiln pulverizing the resultant clinker, as defined in ASTM C150. Portland cement is also considered a hydraulic cement, because it must be mixed with water to form a cement-water paste with the ability to develop strength and harden, even under water.

Cemented soil. Soil in which the grains or aggregates adhere firmly and are bound together. Examples of cementing agents are colloidal clay, iron, silica, alumina hydrate, or calcium carbonate.

Cementing. Remediation technique in which wastes are stirred with water and mixed with cement.

Cenozoic. A geologic era that began about 66.4 million years ago and is composed of the Quaternary and Tertiary periods.

Centigrade (Celsius). Temperature grade used by scientists in which the freezing point of water is represented by 0 and the boiling point by 100. To convert from Celsius to Fahrenheit, multiply degrees Celsius by 9/5 and add 32 to the product.

Centralizer. Device that assists in centering tubular materials in a borehole.

Centrifugal pump. A pump consisting of an impeller fixed on a rotating shaft, enclosed in a casing, and having an inlet and a discharge connection. The rotating impeller creates pressure in the liquid by the velocity derived from centrifugal force.

Centrifugal screw pump. A centrifugal pump having a screw-type impeller.

Centrifugation. The separation of components of a liquid colloidal system using centrifugal force.

Centrifuge. Mechanical device in which centrifugal force is used to separate solids from liquids and/or to separate liquids of different densities.

CERCLA. Comprehensive Environmental Response, Compensation, and Liability Act of 1980 (Superfund); a federal law authorizing identification and remediation of abandoned hazardous waste sites.

Ceramic. A substance composed primarily of clay and similar nonmetallic materials which is formed at high temperatures.

Cetane number. A number that indicates the ignition quality of diesel fuel as a percentage by volume of centane (i.e., hexadecane) in a reference mixture having the same ignition characteristics as the sample fuel. It is analogous to the octane rating scale for gasoline.

CFR. Code of Federal Regulations. A codification of rules published in the Federal Register by the executive departments and agencies of the federal government. The code is divided into 50 Titles representing broad areas subject to federal regulations.

Chalk. A fine limestone formed of minute fossil fragments.

Chamber. (1) In soil micromorphology, pores that are connected with channels; (2) any space enclosed by walls.

Chamber pressure filters. A filtration technique where a group of cloth-covered plates compresses the waste, with the filtrate exiting through the cloth.

Channel. A natural or artificial waterway which periodically or continuously contains moving water or which forms a connecting link between two bodies of water.

Channel accretion. The gradual building up of a channel bottom or bank as a result of sediment deposition.

Channel inflow. Water that flows into a channel system from surface flow, subsurface flow, base flow, or rainfall directly on the channel.

Channel loss. The loss of water from a channel by capillary action and percolation.

Channelization. The modification of a stream channel, such as deepening or straightening, usually with the objective of reducing flooding.

Characteristic waste. A solid waste that is a hazardous waste because it exhibits one or more of the following hazardous characteristics: ignitability, corrosivity, reactivity, or toxicity (*see also* toxicity characteristic rule and toxicity characteristic leaching procedure).

Check gate. A gate consisting of a plank set in a cutoff wall of an irrigation ditch to permit water delivery from the ditch to adjacent land or to other ditches.

Chelating agent. An organic compound that can incorporate metal ions into its structure.

Chelation. The formation of an inner complex compound soluble in water in which the same molecule is attached to a central atom at two different points, forming a ring structure.

Chemical Abstract Services Number (CAS No.). A classification system that provides universal identification of a chemical.

Chemical analysis. Analysis by chemical methods of a substance that shows the composition and concentration of the substance.

Chemical coagulation. Destabilization and initial aggregation of colloidal and finely divided suspended matter by the addition of a floc-forming chemical.

Chemical equivalent. The weight in grams of a substance that combines with or displaces 1 gram of hydrogen. It is obtained by dividing the formula weight by its valence.

Chemical fixation/stabilization. Chemicals and reagent/additives used to cross-link with chemicals to produce a solid matrix.

Chemical kinetics. The study of the speed or velocity of reactions.

Chemical oxygen demand (COD). The results of a laboratory chemical analytical technique used to measure the amount of oxygen required to oxidize all compounds in an organic or water sample (dichromate is the oxidizing agent used in the analysis).

Chemical precipitation. Precipitation induced by the addition of chemicals.

Chemical sludge. Sludge obtained by the treatment of wastewater with chemicals.

Chemical treatment. (1) Process that alters the chemical structure of the hazardous constituents so that resulting material is less hazardous than the original waste; (2) any process involving the addition of chemicals to obtain a desired result.

Chemisorption. Phenomenon related to adsorption in which atoms or molecules of a reacting substance are held to the surface atoms of a catalyst by electrostatic forces that have about the same strength as chemical bonds. Chemisorption is distinguished from physical adsorption by the strength of the bonding, which is less in adsorption.

Chemistry. The science concerned with the nature, properties, and composition of substances and with the development of laws and theories that interpret chemical phenomena in a logical manner.

Chemotrophs. Organisms that obtain energy from oxidation or reduction of inorganic or organic matter.

Chi-square goodness of fit statistic or test. A means to calculate a chi-square statistic that compares observed to expected frequencies in a data set.

Chi-square statistic. The sum of observed minus expected frequencies squared and divided by the expected value. It is used to compare observed to expected data frequencies.

Chi-square test. A statistical test used to determine whether a set of data properly fits a specified distribution within a specified probability.

Chloramines. Compounds of organic or inorganic nitrogen and chlorine.

Chlordane ($C_{10}H_6Cl_8$). A contact insecticide. The emulsifiable formulation concentrate of chlordane alone or in combination with heptachlor is used exclusively in the U.S. for subterranean termite control applications.

Chlorinated hydrocarbon. A group of synthetic organic chemicals produced by replacing one or more of the hydrogen atoms with chlorine. Examples are trichloroethylene, chloroform, tetrachloride, and DDT.

Chlorinated polyethylene. A family of polymers produced by a chemical reaction of chlorine on polyethylene.

Chlorination chamber. A detention basin designed to secure the diffusion of chlorine through a liquid.

Chlorinolysis. Chemical treatment process where excess chlorine is added to waste at high temperatures rendering it less hazardous.

Chloroform ($CHCl_3$). A volatile narcotic liquid produced by reacting acetone with chlorinated lime or by the chlorination of methane.

Chlorohydrin ($CH_2OHCHOHCH_2Cl$). A group of aliphatic compounds that contain one or more hydroxyl groups and at least one chlorine atom. It is made by reacting glycerol with hydrogen chloride.

Chloroprene. A chemical used in the production of synthetic rubber.

Chlorosulfonated polyethylene. A family of polymers produced by the reaction of polyethylene with chlorine and sulfur dioxide.

Cholinesterase. An enzyme that helps regulate nerve impulses. Cholinesterase inhibition is associated with a variety of acute symptoms such as nausea, vomiting, blurred vision, stomach cramps, and rapid heart rate and can lead to death in severe cases.

Chroma. In soil morphology, the measure of the purity or strength of spectral color.

Chromatographic column. A column consisting of a tube whose tendency to retain or pass a compound carried into it by a carrier gas stream will vary depending on properties of the compound. The tube may be glass, stainless steel or Teflon®. Two types of analytical columns exist — packed columns and capillary columns.

Chromatography. A process of fractioning or separating components of gaseous or liquid mixtures. The method involves passing a gaseous or liquid mixture through a column of porous material often coated with a nonvolatile liquid called the stationary phase. The components of the mixture are adsorbed selectively and pass through the porous material at different rates, emerging as distinct separation zones at the terminus of the column. A carrier gas or liquid is used to provide movement of the material through the adsorbent. When a compound has been separated and detected, the identity of each component can be determined.

Chromatography, gas (GC). An analytical technique based on the vaporization of a liquid sample followed by the separation of the gaseous components that are then identified and quantified.

Chromel®. One of several nickel-based alloys which are resistant to oxidation and loss of strength due to heat and have a high electrical resistivity and low temperature resistance coefficient.

Chromic acid (CrO₃). Water-soluble inorganic acid that is a strong oxidizing agent and highly corrosive; commonly used as a colorant in ceramic products, metal cleaning, and metal plating.

Chronic toxicity. The property of causing death or damage to an organism by poisoning during prolonged exposure, which, depending on the organism tested and the test conditions, may range from several days to weeks, months, or years.

Circulating bed combustion. A variation of fluidized bed technology which does not employ a fixed bed depth.

Cis-. A chemical prefix that means "on this side" and is the opposite of *trans-*, which means "on the other side" or "beyond". The prefix is used to indicate the position of substituent atoms or groups in relation to double-bonded carbons.

Cladding. A metallurgical term used to designate a type of protective coating consisting of a relatively thick layer of a metal, such as nickel, mechanically applied to a substrate metal, such as steel or copper. Clad metals are used in the electrical field for semiconductors and corrosion resistant equipment.

Clarification. Any process or combination of processes designed to reduce the concentration of suspended matter in a liquid.

Clarifier. A mechanical device used for removing solids from water.

Class. A group of units similar in selected properties and distinguished from all other classes of the same population by differences in these properties.

Clastic. Pertaining to rock or sediment composed principally of broken fragments that are derived from pre-existing rocks or minerals and that have been transported some distance from their places of origin.

Clay. (1) Soil fraction whose particles range from 2 microns (μm) downward; (2) in the Unified Soil Classification System, a soil that contains more than 50% particles with diameters less than 0.047 millimeters (mm). Clay includes the colloidal fraction, and most clays belong to a group of minerals known as aluminosilicates.

Clay cutan (argillan). Clay that has been transported by water through the larger pores and deposited on the walls after the water has drained from the pores.

Clay loam. Soil containing 20 to 45% sand and 27 to 40% clay.

Claypan. A stratum or horizon of accumulated stiff, compact, and relatively impervious clay.

Clean Air Act (CAA). Law that authorizes regulations governing releases of airborne contaminants from stationary and nonstationary sources. The regulations include National Ambient Air Quality Standards for specific pollutants.

Clean Water Act (CWA). Law that authorizes establishment of the regulatory program to restore and maintain the physical and biological integrity of the nation's waters. The CWA established, among other things, the National Pollutant Discharge Elimination System (NPDES) to regulate industrial and municipal point-source discharges.

Cleanroom. The super-clean environment in which semiconductors are manufactured. The smaller the "class" (i.e., Class 1 vs. Class 100), the cleaner the facility. The class number relates to the number of particles smaller than 0.10 micron (μm) in size that can be found per cubic foot of air in the cleanroom.

Cleavage. The capacity of a material to split.

Clinometer. An instrument used to measure the dip or inclination of a surface.

Closed portion. The portion of a facility that an owner or operator has closed in accordance with the approved facility closure plan and all applicable closure requirements.

Closure plan. A written plan (subject to approval by authorized regulatory agencies) which the owner/operator of a hazardous waste management facility must submit with the RCRA permit application or for interim status closure. The approved plan becomes part of the permit conditions subsequently imposed on the applicant. The plan identifies steps required to (1) close a hazardous waste management unit completely or partially at any point during its intended operating life, and (2) close the unit completely at the end of its intended operating life.

Cluster analysis. A procedure that groups observations from a multivariate data set into clusters of observations points.

Coagulant. Compound responsible for coagulation; a floc-forming agent.

Coagulation. Destabilization and initial aggregation of colloidal and finely divided suspended matter by the addition of a floc-forming chemical or by biological processes.

Coal. A carbonaceous solid material formed throughout several geological eras from vegetation by pressure, heat, and bacterial actions out of contact with air.

Coal gas. General term for any number of a variety of gases manufactured from coal.

Coal tar. A thick liquid obtained as a byproduct of coal distillation; the major product of this distillation is coke. Organic compounds derived from coal tar include benzene, naphthalene, phenol cresol, and anthracene. Approximately 9 gallons of coal tar are obtained from 1 ton of coal.

Cobble. A rock having a diameter between 3 and 8 inches.

Coefficient. A numerical quantity interposed in a formula that expresses the relationship between two or more variables.

Coefficient of permeability. An obsolete term that has been replaced by the term *hydraulic conductivity*.

Coefficient of storage (S). Volume of water released from or taken into storage per unit surface area of an aquifer per unit change in the component of head normal to that surface. (S) is a dimensionless index and ranges form 3.0×10^{-1} to 1.0×10^{-5}.

Coefficient of transmissivity. *See* transmissivity.

Coefficient of viscosity. Numerical factor that measures the force required to maintain a unit difference in velocity between two layers of water a unit distance apart.

Cohesion. The force of molecular attraction between the particles of any substance which tends to hold them together

Coke. The end product of the distillation of bituminous coal or the cracking of petroleum hydrocarbons. Coal-derived coke is used as a fuel for blast furnaces, while petroleum-derived coke is used to make electrolytic and electrothermic electrodes.

Coke bronze. The fines (roughly smaller than 2 centimeters) which are separated from the coke at the coke screening station. Typically, 3 to 8% of the coal charged ends up as recovered coke breeze, which is often used for power generation.

Coke oven gas. A gas rich in hydrogen and methane (heating value of about 530 Btu per cubic foot) made by carbonizing coal in byproduct coke ovens, which are comprised of batteries of long, narrow, refractory-lined ovens where coal is heated in the absence of air.

Collimator. Device used to develop a beam of limited cross-sections of molecules, atoms, or nuclear particles in which the paths of the particles are parallel.

Colloid. Extremely small solid particles, 0.0001 to 1 micron in size, that will not settle out of a solution. A colloid is intermediate between a true dissolved particle and a suspended solid that will settle out of solution.

Colloid chemistry. Study of the dispersion in solids, liquids, or gases when one dimension is less than 1 micron but greater than 1 millimicron.

Colloid mill. Device used to grind coarse materials into colloidal-sized particles.

Colluvial. Consisting of alluvium and of angular fragments of the original rocks.

Colluvial deposit. A heterogeneous deposit of rock waste, such as talus, cliff, and avalanche accumulations, which results from the transporting action of gravity.

Colluvium. A deposit of rock fragments and soil material that has accumulated at the base of steep slopes as a result of gravity.

Colorimetry. Measurement of color naturally present in samples or developed therein by the addition of reagents.

Cometabolism. The simultaneous metabolism of two compounds, in which the degradation of the second compound (the secondary substrate) depends on the presence of the first compound (the primary substrate). For example, in the process of degrading methane, some bacteria can degrade hazardous chlorinated solvents that they would otherwise be unable to attack.

Complementary metal oxide semiconductor (CMOS). A MOS device containing both N-channel and P-channel MOS active elements. One of two basic processes (MOS and bipolar) used to fabricate integrated circuits.

Complex ions. Soluble species that are formed through the combination of other, simpler, positively charged species in the solution.

Complexation. Reaction in which a metal ion and one or more anionic ligands chemically bond. Complexes often prevent the precipitation of metals.

Composite. Any combination or blend of solid materials whose particles or fibers are large enough to be visible to the naked eye and which are chemically different.

Compositing. A sampling technique in which grab samples are collected and composited into one sample, and the composited sample is analyzed.

Composting. The controlled biodegradation of the organic portion of a solid waste pile, such as food waste, grass clippings, and leaves.

Compound alluvial fan. Series of alluvial fans which merge into one other.

Compound pipe. Pipeline made up of two or more pipes of different diameters.

Compressibility. Fundamental material property that describes a change in volume or strain induced in a material under an applied stress.

Concentration. The quantity of any substance (solid, liquid, or gaseous) present in a specified weight or volume of a mixture.

Condensate. The liquid that separates from a vapor during condensation.

Condensation. The process by which a substance changes from the vapor to the liquid or the solid state.

Conditionally exempt small quantity generator (CESQG). Those who generate no more than 100 kilograms of hazardous waste per month. Other than the hazardous waste determination requirement in 40 CFR 262.11, CESQGs are exempt from RCRAs, provided they do not exceed certain quantity limits for hazardous waste storage or generation.

Conductance. The reciprocal of resistance as expressed in reciprocal ohms or mhos.

Conductance, specific. A measure of the ability of water to conduct electric current at 77°F. It is related to the total concentration of ions in the water.

Conduction. The transfer of heat between two parts of a stationary system caused by a temperature difference between the parts.

Conductimetry. An electrical method of analysis used to measure the ability of a solution to transmit an electrical current.

Conductivity, solution. Measure of the ability of a liquid to carry an electrical current. The conductivity of a solution varies with both the number and type of ions in the solution.

Cone of depression. A depression in the groundwater table or potentiometric surface that has the shape of an inverted cone and develops around a well from which water is withdrawn. It defines the area of influence of a well.

Cone of influence. A depression, roughly conical in shape, produced in a water table or other piezometric surface by the water extraction from a well at a given rate.

Cone penetrometer test (CPT). Device that measures the lithology in unconsolidated deposits. The cone penetrometer consists of a 60° apex cone tip and cylindrical friction sleeve that are attached to strain-gauged load cells that measure the soil's resistance to penetration. The penetrometer measures the tip resistance, sleeve friction, pore pressure of the soils, and inclination as it is hydraulically driven into the subsurface.

Confidence interval. The random interval constructed from sample data in such a way that the probability that the interval will contain the true value of the parameter is a specified value.

Confidence level (confidence limit, 95%). A level of data reliability achieved by setting a percent confidence limit. A 95% confidence limit is the limit of the range of analytical values within which a single analysis will be included 95% of the time.

Confined groundwater. Groundwater that is under pressure greater than atmospheric; its upper limit is the bottom of a bed of distinctly lower hydraulic conductivity than that of the material in which the confined water occurs.

Confined water well. A well whose sole source of supply is confined groundwater.

Confining bed. A term synonymous with *aquiclude* and *aquitard* in many reports. It generally refers to a body of "impermeable" material stratigraphically adjacent to one or more aquifers that restricts the movement of groundwater into or out of adjacent aquifers.

Confining layer. An aquitard or impermeable layer that confines the limits of an aquifer.

Confluent stream. A stream that unites with another stream.

Conjugate acid. Solution of weak bases and their salts.

Conjugate base. Solution of weak acids and their salts.

Conjugated compound. Organic compound which contains two (or more) double bonds with a single bond positioned between them.

Conjugation. In microbiology, the direct transfer of chromosomes between attached cells.

Connate water. Water entrapped in the interstices of a sedimentary rock at the time it was deposited.

Consecutive reaction. Complex chemical reaction in which products of one reaction become the reactants of a following reaction.

Conservation of energy law. Law of thermodynamics that states that for a given system energy can be absorbed from, or released to, the outside; along the way, it can change form but can be neither created or destroyed.

Conservative. A contaminant that does not degrade and the movement of which is not retarded; it is unreactive.

Consistency, soil. The degree of adhesion between soil particles that can resist deformation or rupture.

Consolidated rocks. Mineral particles of different shapes and sizes that have been integrated into a solid mass by heat, pressure, or chemical processes.

Consolidation. In soils, the process by which the soil is compacted or compressed in a manner that results in a reduction of pore volume due to the exclusion of water.

Constatan. Alloy consisting of 55% copper and 45% nickel used for electrical resistance heating and thermocouples.

Constituent. An essential part or component of a system or group (e.g., an ingredient of a chemical mixture). For instance, benzene is one constituent of gasoline.

Consumptive use of water. Any use of water which depletes the available supply.

Contact. A boundary surface between two rock units, especially between an intrusive and its host rock.

Contact filter. A filter used in a water treatment plant for the partial removal of turbidity before final filtration.

Contact metamorphism. Metamorphic changes in rock due to the proximity of an intrusion into the host rock.

Contour. Line of equal value above a specified datum.

Contour interval. Difference in value between adjacent contours on a map.

Contour line. A line joining points having or representing equal values.

Contour map. Map showing the configuration of the surface by means of contour lines drawn at regular intervals of value.

Control flume. A flume arranged for measuring the flow of water, wastewater, or other liquid.

Convection. In physics, the mass motions within a fluid resulting in transport and mixing of fluid properties; caused by the force of gravity and by differences in density due to nonuniform temperature.

Cooling tower. A hollow, vertical structure with internal baffles that break up falling water so that it is cooled by upward-flowing air and by water evaporation.

Coordination. In chemistry, the formation of a weak bond between organic molecules, which are capable of donating electrons, and adsorbed cations, which are capable of accepting electrons.

Coordination compound. A molecule (often called a complex), either charged or neutral, formed by the attachment of a transition-metal ion to another molecule or ion by means of a coordinate covalent bond (i.e., a covalent bond in which both the shared electrons are furnished by the same atom).

Coordination number. A reference to the number of ligands attached to a central ion.

Copolymer. A high-polymer substance, usually an elastomer, made up of two or more different kinds of monomer (for example, styrene and butadiene).

Copper (Cu). A ductile, malleable reddish-brown metallic element that is toxic to aquatic organisms, from algae and plants to fish.

Copper sulfate. Chemical prepared from copper and sulfuric acid.

Copperas ($FeSO_47H_2O$). Common name for ferrous sulfate heptahydrate.

Core drill. An instrument for boring holes in rock or other material to obtain a cylindrical sample of the material.

Core drilling. A method of drilling with a hollow bit and core barrel to obtain a rock core.

Corrective action management unit (CAMU). An area within a facility that is broadly contaminated by hazardous wastes and which contains discrete land-based units.

Corrective Action Reporting System (CARS). EPA's national database of information on corrective action permits and enforcement actions.

Corrective Action Rule (also known as Subpart S). The proposed regulation (see the *Federal Register,* July 27, 1990) which would define requirements for corrective action under RCRA corrective action permits (3004(u)). Although the proposed rule establishes regulations for the corrective action permit program, the EPA also intends use of these regulations as guidance for corrective action under enforcement (3008(h)).

Corrective measures implementation (CMI). The fourth and final step in maintaining and monitoring selected corrective measures that have been approved by the regulatory agency. This stage combines activities that are often segregated and corrective action, RCRA facility assessment, RCRA facility investigation, and corrective measures study.

Corrective measures study (CMS). The third step in the RCRA corrective action process. If the RCRA facility investigation (RFI) reveals a potential need for corrective measures, the agency requires the owner to perform a CMS to identify and recommend specific measures to correct the releases. Although analogous to the Superfund feasibility study (FS) stage, this study is usually less complicated.

Corrosion. An electrochemical change in a metal surface caused by the reaction of the metal with one or more substances with which it is in contact for long periods. This change results in the gradual deterioration of the substance.

Corrosivity. Characteristic of a hazardous waste whereby it dissolves metals or burns the skin.

Coulomb. A meter/kilogram/second unit of quantity of electricity equal to the quantity of charge transferred in 1 second across a conductor in which there is a constant current of 1 ampere.

Coulometric titration. Titration method that measures the quantity of electricity passed during an electron exchange involving the substance being determined.

Country rock. A rock with an igneous intrusion.

Coupling. (1) Chemical reaction taking place during the union of amino acids to form proteins; (2) polymerization of certain phenols (especially those containing two substituted methyl groups) by means of oxygen and a nitrogenous catalyst to form a thermoplastic polymer; (3) a low molecular weight group that forms a linkage in block polymers; (4) in the chemistry of dyeing, a reaction between an electron donor, such as a phenol or arylamine, and an electron-accepting diazonium compound to form an azo dye.

Covalent bond. Type of chemical bond in which atoms of the same or different elements combine to form a molecule (or in some cases a crystal) by sharing pairs of electrons.

Covariance. Measurement of the linear association between two variables. If both variables fall above or below their means at the same time, a positive covariance results. If one variable is above its mean while the other is below, the covariance is negative.

Cresol. Organic liquid derived from coal tar or by reacting methyl alcohol with phenol.

Crest gate. A gate installed on the crest of the spillway of a dam, which is operated to vary the discharge over the spillway.

Cretaceous. A geologic period between 144 and 66.4 million years ago.

Crib weir. A low weir built of log cribs filled with rock.

Cross-linking. The union or binding together of two or more polymer chains by either (1) heating in the presence of a substance capable of forming a chemical bond between the two chains, or (2) exposing the polymer to ionizing radiation.

Cryic soil. A soil with a mean annual soil temperature between 0 to 8°C with summer temperatures less than 15°C.

Cubic meter. 1 cubic meter = 35.3 cubic feet.

Cubic spline. Method to smooth a grid. Cubic splines are used to connect a set of points with a smooth curve. This technique simulates the manual drafting technique of using a flexible strip of metal or plastic to draw a smooth curve between data points.

Cumulative distribution function. Statistical term synonymous with the distribution function.

Cumulative objective. A numerical water quality objective limiting the total concentration of a group of constituents regardless of the characteristics of the individual members of the group.

Curie (Ci). A unit of measurement of radioactivity equal to 3.7×10^{10} disintegrations per second.

Cutan. In soil morphology, the modification of the texture, structure, or fabric in soil materials due to the concentration of particular soil constituents or *in situ* modification of the surface. Categories of cutans are clay, stress, oxide, and organic matter.

Cutoff trench. Trench excavated below the normal base of a dam or other structure and filled with relatively impervious material to reduce percolation under the structure.

Cutoff wall. A thin wall or footing constructed downward from, under, or around the headwall and lip wall of a dam that provides resistance to seepage.

Cyanogen. Highly toxic gas used as a starting material for the production of complex thiocyanates used as insecticides.

Cyclic amide. An amide situated in a ring of carbon atoms.

Cyclic anhydride. A ring compound formed by the removal of water from a compound.

Cyclic compound. An organic compound containing one or more closed rings. If the ring contains only carbon atoms, it is called carbocyclic and is represented by alicyclic compounds (cyclohexane) and aromatics (benzene); however, if one or more atoms in the ring is an element rather than carbon (e.g., nitrogen, sulfur, etc.), the compound is called heterocyclic.

Cyclization. Rearranging open-chain hydrocarbons to a closed ring.

Cycloaddition. A chemical reaction where unsaturated molecules combine to form a cyclic compound.

Cycloaddition reactions. In chemistry, reactions where two more unsaturated groups combine to generate a cyclic structure containing fewer bonds than the reactants.

Cyclobutadiene. A cyclic compound containing two alternate double bonds.

Cyclobutane. An alicyclic hydrocarbon used in organic synthesis.

Cyclohexane (C_6H_{12}). A saturated alicyclic compound that occurs in petroleum and can be made by hydrogenating benzene with the aid of a catalyst.

Cyclohexanone. An oily liquid tone that is soluble in alcohol, ether, and other organic solvents. It is used as an industrial solvent in the production of adipic acid and in the preparation of cyclohexanone resins.

Cyclone dust separator. A device used to remove dust particles from air.

Cyclopentane. A cyclic hydrocarbon used to improve anti-knock and combustion properties of gasoline.

Cyclopentene. A liquid used as a chemical intermediate in the petroleum chemistry.

D

Dacthal. A selective pre-emergence herbicide

Dalton's law of partial pressure. Law that states that the total pressure of a mixture of gases equals the sum of the partial pressures.

Darcy. Unit of measurement commonly used in the petroleum industry to express intrinsic permeability; equal to 9.87×10^{-9} cm^2.

Darcy's Law. An equation formulated in 1856 by Henry Darcy from his work of water flow through sand filter beds. The formula is used to describe fluid flow through a porous media. One expression is

$$Q = -KiA$$

where

    Q = rate of inflow into the porous media.
    i = the hydraulic gradient.
    A = area.

This formula assumes that the rate of viscous flow of water in isotropic porous media is proportional to, and in the direction of, the hydraulic gradient. The rate of viscous flow of homogenous fluids through isotropic porous media is also assumed to be proportional to, and in the direction of, the driving force.

Datum. An agreed standard point or plane of stated elevation.

Datum plane. A surface used as a reference from which to compute height or depths.

Daughter product. The immediate product of radioactive decay of an element.

DDE (dichlorodiphenyldichloroethylene). A product of degradation of DDT resulting from the loss of one molecule of hydrochloric acid (dehydrohalogenation). DDE degrades to DDA by the loss of two more molecules of hydrochloric acid (HCl).

DDT (dichlorodiphenyltrichloroethane). An organic pesticide used as a delousing agent and as an insecticide to control the spread of diseases such as malaria, yellow fever, and typhus. A persistent insecticide which is a mixture of isomers of dichlorodiphenyltrichloroethane (a chlorinated hydrocarbon) having the formula $(ClC_6H_4)_2CHCCl_3$. It is derived from chloral and chlorobenzene by a condensation reaction. It has a half-life of 15 years and can collect in fatty tissues of certain animals. The EPA banned registration and interstate sale of DDT for virtually all but emergency uses in the U.S. in 1972 because of its persistence in the environment and accumulation in the food chain.

Deacidify. The process of reducing acidity.

Dealkalization (solodization). The removal of sodium ions from the exchange sites of a soil.

Dealkylate. To remove the alkyl groups from a compound.

Debenzyolation. Removal of a molecule from the benzyl group.

Debris. Any material, including floating trash, suspended sediment, or bed load, moved by a flowing stream.

Debye-Huckel theory. A theory of the behavior of strong electrolytes, according to which each ion is surrounded by an ionic atmosphere of a charge of the opposite sign whose behavior retards the movement of ions when a current is passed through the medium.

Debye relaxation time. Time required for the ionic atmosphere of a charge to reach equilibrium in a current-carrying electrolyte.

Decalcification. A reaction that removes calcium carbonate from one or more soil horizons.

Decane ($CH_3(CH_2)_8CH_3$). A moderately flammable liquid hydrocarbon obtained from petroleum refining.

Decantation. Separation of a liquid from solids, or from a liquid of higher density, by drawing off the upper layer after the heavier material has settled.

Decay. The spontaneous conversion of a portion of the mass of a natural or artificial radioactive element or nuclide into energy in the form of alpha, beta, and gamma radiation. This results from an unstable nuclear structure.

Dechlorination. A chemical treatment process where chlorine is chemically removed from a chlorinated organic compound.

Decibels A-scale (db). A unit of frequency-weighted noise used in traffic and industrial monitoring. Decibel A-scale corresponds to the frequency response of the ear.

Decomposition. The structural breakdown of a molecule into simpler molecules or atoms.

Decomposition potential. The electrode potential at which an electrolysis current begins to increase.

Deep percolation. The moisture or water that passes below the root zone of plants.

Deferrization. Removal of soluble compounds of iron from water.

Deflocculator. A dispersing agent used to retard settling of solid particles in a suspension, especially when the particles tend to clump together and settle out rapidly. Emulsifiers are often effective deflocculators.

Defoaming agent. A material having low compatibility with foam and a low surface tension.

Defoliant. A chemical or material applied to plants that causes them to shed their leaves.

Degradation. The chemical or biological conversion of a complex compound into simpler compounds.

Degradation potential. The degree to which a substance is likely to be reduced to a simpler form by bacterial activity.

Degreasing. The process of removing greases and oils from waste, wastewater, sludge, or solid wastes; the industrial process of removing grease and oils from machine parts or iron products.

Degree of freedom. A variable including pressure, temperature, composition, and specific volume that is specified to define the state of a system.

Degree of polymerization (DP). A number indicating the extent to which the molecules of a monomer have combined to form a polymer; it is determined by calculating the number of such molecules present in an average molecule of polymer in a sample.

Degree of saturation (S or SR). A ratio that expresses the volume of water present in the soil relative to the volume of pores. Values of zero and 100% indicate a dry and a saturated soil, respectively.

Dehydration. A chemical or physical process whereby water in chemical or physical combination with other matter is removed.

Dehydrator. An agent that removes water from a material.

Dehydrochlorination. A process whereby a molecule of hydrogen chloride is removed from an organic chloride leaving a double or triple bond in the organic compound. It is the principal process for manufacturing vinyl chloride from ethylene dichloride.

Dehydrocyclization. A catalytic petroleum reforming reaction by which straight-chain paraffin hydrocarbons containing up to five carbon atoms are dehydrogenated to unsaturated structures, which are then converted into aromatic (ring) compounds at 400°C (about 750°F). This greatly increases the octane number and is one important method of making high-quality motor fuels.

Dehydrogenation. The removal of hydrogen from a compound.

Dehydrohalogenation (dehydrodehalogenation). Chemical removal of hydrogen and a halogen from a compound.

Deionization. An ion-exchange process in which the charged species or ionizable organic and inorganic salts are removed from solution.

Deliquescent. Term used to characterize water-soluble salts, usually in finely divided form (small plates, crystals), which not only absorb moisture from the air, but also tend to soften and even dissolve as a result of this absorption.

Demal. A unit equal to the concentration of a solution in which 1 gram equivalent of solute is dissolved in 1 cubic decimeter of solvent.

Demethylation. The removal of the methyl group from a compound.

Denaturant. A substance added to ethyl alcohol to prevent its being used for internal consumption. This applies chiefly to ethyl alcohol intended for industrial use, the contaminating material being added to ensure that it is not diverted to beverage use.

Denature. The change of a protein so that the original properties such as solubility are changed as a result of the protein's molecular structure.

Denier. Term used in the textile industry to designate the weight per unit length of a filament (i.e., its diameter or "fineness").

Denitration. The denitrification removal of nitrates or nitrogen.

Denitrification. Bacterial reduction of nitrite to gaseous nitrogen under anaerobic conditions.

Dense hydrocarbon. Petroleum products with specific gravities greater than that of water.

Density. The amount of mass per unit volume.

Density gradient centrifugation. Separation of particles according to density.

Density of mercury. A physical property equal to 13.5×10^3 kilograms/meter^3 at 20°C.

Density of solids (ρ_s). The ratio of the density of a material to water at 4°C and at atmospheric pressure. The term is synonymous with mean particle density and specific gravity. For mineral soils, the mean density of the particles ranges from 2.6 to 2.7 g/cm^3.

Density of stratification. The formation of identifiable layers of different densities in bodies of water.

Density of water (0°C). A physical property equal to 1×10^3 kg/m^3.

Deoxidizer. A substance that reduces the amount of oxygen in a substance.

Deoxygenation. The removal of oxygen from a substance.

Depolymerization. Decomposition of macromoleculars into simple compounds.

Deposition. Geologic process involving the accumulation of rock material or other debris.

Derivative. A substance made from another substance.

Desiccant. Material used to reduce the relative humidity of a substance or a container to near zero humidity. Anhydrous calcium chloride is a desiccant.

Desiccation. Removal of water vapor from a material by a hygroscopic substance, such as calcium chloride or silica gel, placed in an air-tight container with the material to be dried (often under vacuum).

Desiccator. A sealed container in which the humidity is kept near 0%.

Desilication (ferrallitization, ferritization, allitization). The chemical migration of silica from a soil horizon.

Desorption. The process of removing a sorbed substance. Desorption is the reverse of adsorption or absorption.

Destructive distillation. Decomposition of organic compounds by heat but without air.

Desulfonate. The removal of a sulfonated molecule from a group.

Detoxifier. An adaptation of drilling technology that allows the use of a range of *in situ* treatment methods such as air-/stream-stripping, neutralization, solidification/stabilization, and oxidation.

Detritus. The heavier mineral debris moved by natural watercourses, usually in bed-load form.

Developing agent. (1) In photographic chemistry, a reducing compound which reacts with silver bromide crystals to form metallic silver under the influence of light, which acts as the catalyst; (2) in dyeing technology, an organic compound which unites or combines with another compound in or on the fiber to form or "develop" a new color, often having improved properties.

Devitrification. A process by which the glassy texture of a material is converted into a crystalline texture.

Dialdehyde. A molecule that has two aldehyde groups.

Dialysis. The selective diffusion of a molecule through a membrane.

Diamide. A molecule that has two amides.

Diamine. A compound containing two amino groups.

Diamyl phenol. A liquid used in synthetic resins with a boiling range of 280 to 295°F.

Diazine. A hydrocarbon that contains an unsaturated hexatomic ring of two nitrogen atoms of an alkane molecule that has been replaced by a diazo group.

Diazotization. A chemical reaction used in the production of azo dyes and a wide range of organic intermediates.

Diborane. A highly flammable and explosive gas composed of boron and hydrogen (B_2H_6). Diborane will ignite at a temperature of 38°C (100°F).

Dibromochloropropane. A liquid used as a nematicide for crops.

Dibromodifluoromethane. A liquid with a boiling point of 24.5°C; it is soluble in methanol and ether.

Dibutyl. A compound with two butyl groupings bonded through a third atom or group in a molecule.

Dibutyl phthalate ($C_6H_4(Co_2C_4H_9)_2$). A slightly viscous liquid, without odor or color that is made by reacting butyl alcohol with phthatic anhydride. It is a liquid commonly used as a plasticizer and insect repellent.

Dibutyl succinate. A liquid that is insoluble in water and is used as an insect repellent for cattle flies, cockroaches, and ants around barns.

Dicalcium. A molecule that contains two atoms of calcium.

Dichloride. An inorganic salt or organic compound that has two chloride atoms in its molecule.

Dichlorobenzene. A chlorinated aromatic hydrocarbon ($C_6H_4Cl_2$), having three isometric forms, depending on the positions at which the chlorine atoms are attached to the benzene nucleus.

Dichlorodifluoromethane ($CHCl_2F$). A fluorocarbon gas created by the catalytic reaction of hydrogen fluoride and carbon tetrachloride.

Dichloroethane (1,2-DCA). A chlorinated hydrocarbon used as a lead scavenger in the 1960s and 1970s in unleaded gasoline. It is also used as a solvent.

Dichloroethene (1,1-DCE; $C_2H_2Cl_2$). Also known as vinylidene chloride; a volatile, colorless liquid that polymerizes easily and has a mild, sweet odor.

Dichloroethyl ether. A liquid commonly used as a solvent in paints, varnishes, and lacquers and as a soil fumigant.

Dichlorofluoromethane. A heavy gas used in fire extinguishers and as a solvent, refrigerant, and aerosol propellant.

Dichloropentane. A light-yellow liquid used as solvent, paint and varnish remover, insecticide, and soil fumigant.

Dichroic. Term used in crystallography to denote crystals which refract incident light in two directions, thus displaying two colors when observed from different angles.

Dichromate. A salt of dichromic acid.

Die. A single rectangular piece of semiconductor material onto which specific electrical circuits have been fabricated; refers to a semiconductor which has not yet been packaged.

Dieldrin. A white insecticide used in mothproofing carpets and other furniture. An insecticide used for control of soil insects, public health insects, termites, and other pest. Except for termite control, the use of dielrin is prohibited in the U.S.

Dielectric constant (permittivity) (ε). A value that serves as an index of the ability of a substance to resist the transmission of an electrostatic force from one charged body to another, as in a condenser. It is described by Coulomb's Law:

$$F = (q_1 q_2)/4\pi\varepsilon r^2$$

where

F = force between two charges.
q_1 and q_2 = magnitude of the two charges.
r = distance between the two charges.
ε = permittivity.

Diesel fuel. A petroleum distillate, either straight-run or partially cracked, that is used as a power source for diesel engines; it is equivalent to the no. 2 grade of domestic fuel oil. It is a complex mixture of hydrocarbons in the carbon range of C_9 to C_{20}. Constituents include n-alkanes, iso- and cycloalkanes, highly branched isoalkanes, aromatics, and polar compounds. It can also contain small amounts of n-hexane, benzene, toluene, xylenes, and ethylbenzene.

Diesel fuel no. 2. A fuel used in motor vehicles which generally contains straight-run kerosene, straight-run middle distillates, hydrodesulfurized middle distillates, and light catalytically and thermally cracked distillates.

Dimethyl carbinol. An alcohol-soluble liquid used in pharmaceuticals.

Dimethyl carbonate. A colorless liquid soluble with most organic solvents and used as a solvent.

Dimethyl succinate. A water-white liquid used as a chemical intermediate and plasticizer.

Dimethyl sulfate (ethyl sulfate). A colorless oil used as an intermediate in organic synthesis.

Differential settlement. The uneven lowering of different parts of an engineered structure, often resulting in damage to the structure.

Differential spectrophotometry. The spectrophotometric analysis of a sample when a solution of the major component of the sample is placed in the reference cell.

Diffraction. The bending of light around an obstacle.

Diffraction spectrum. Parallel light and dark bands of light produced by diffraction.

Diffuse spectrum. Spectrum with bands that are very broad even when there is no possibility of line broadening by collisions.

Diffuser. Porous plate, tube, or other device through which air is forced and divided into minute bubbles for diffusion in liquids.

Diffusion. The process whereby particles of liquids, gases, or solids intermingle as the result of their spontaneous movement. In dissolved substances, this occurs as the substance moves from a region of higher to lower concentration.

Digestion. Biological decomposition of organic matter in sludge, resulting in partial gasification, liquefaction, and mineralization.

Digestor. A heated pressure vessel where the anaerobic digestion of organic materials occurs.

Dihaloelimination. The reductive elimination of two halide substitutes to form an alkene.

Diisobutyl ketone. A liquid which boils at 168°C; it is soluble in most organic liquids and is toxic and flammable.

Diisocyanate. A compound with two isocyanate groups that is used to produce polyurethane foams, resins, and rubber.

Dilatancy. (1) Increase in the bulk volume of a rock during deformation caused by elastic and nonelastic changes; (2) property of a viscous suspension that solidifies under pressure.

Dilatant. A material that can increase in volume when its shape is changed.

Dilauryl thiodipropionate. White flakes used as an antioxidant, plasticizer, and preservative.

Dilisobutylene. An isomer used in alkylation and as a chemical intermediate.

Diluent. A low-gravity material (solid, liquid, or gas) added to a fuel product either (1) to reduce its cost or (2) to lower the concentration of its base compound for some desirable purpose.

Dilution ratio. A specific term used by lacquer formulators in reference to nitrocellulose solutions; it is also known as the hydrocarbon tolerance of such solutions.

Diluvium. Unsorted and sorted deposits of the glacial period.

Dimer. A molecule resulting from the combination of two identical molecules called monomers.

Dimethrin. An amber liquid that is soluble in petroleum hydrocarbons, alcohols, and methylene chloride which is used as an insecticide for mosquitoes.

Dimethyl. A compound with two methyl groups.

Dimethyl phthalate. An odorless liquid which boils at 282°C; it is soluble in organic solvents, slightly soluble in water.

Dimethylamine. A flammable gas used as an acid-gas absorbent, solvent, and flotation agent in pharmaceutical and electroplating.

2,3-Dimethylbutane. A liquid used as a high-octane fuel.

Dimorphism. Substance that can crystallize in two forms with the same chemical composition.

Dinitrobenzene. A cyclic organic compound consisting of two nitro groups attached to a benzene nucleus.

Dinitrotoluene. One of six isomeric substitution products of benzene used in high explosives; it is formed by nitration of toluene.

Diolefins. An aliphatic compound that contains two double bonds in the molecule.

Dioxin. A chlorinated dibenzodioxin (CDD) and one of the most toxic substances known. It occurs as a byproduct of chemical synthesis, from electrical fires, from combustion of wood preservatives, and from municipal solid waste incinerators. It is one of the 126 priority pollutants listed by the EPA under Section 307(a) of the Clean Water Act.

Dioxygenases. A class of enzymes associated with the transformation of benzene and benzenoid chemicals.

Dip. In geology, the inclination of a geologic surface measured from the horizontal. The dip forms the angle between a line perpendicular to the strike and the horizontal plane.

Direct oxidation. The direct combination of substances with oxidants accomplished without benefit of living organisms.

Dirichlet. In groundwater modeling, a specific head boundary that is established at a model boundary.

Discharge. The volume of water per unit flowing past a fixed point in a river, stream, or pipe.

Discharge area. An area where there is a downward component of the hydraulic head.

Discharge capacity. The maximum rate of flow that a conduit, channel, or other hydraulic structure is capable of passing.

Discharge coefficient. A coefficient by which the theoretical discharge of a fluid through an orifice, weir, nozzle, or other passage must be multiplied to obtain the actual discharge.

Discharge velocity (v_d). For fluid flow through a porous medium, it is equal to the hydraulic gradient times the saturated hydraulic gradient.

Disconformity. A time break in a sequence of beds where the beds above and below the unconformity are parallel but the unconformity is not parallel to the bedding.

Discriminator. A circuit that can be adjusted to accept or reject signals of different amplitude or frequency.

Disinfectant. A substance having the ability to kill or inactivate bacteria.

Dispersion. The spreading and mixing of chemical constituents in groundwater caused by diffusion and mixing due to microscopic variations in velocities within and between pores.

Dispersion force. The attraction between molecules that have no permanent dipole.

Dispersive transport. The spreading of a solute from the path that it would be expected to follow by advective transport as a result of mechanical mixing and molecular diffusion.

Dispersivity. The ability of a contaminant to disperse within the groundwater by molecular diffusion and mechanical mixing.

Displacement. Reaction in which an atom, radical, or molecule displaces and sets free an element of a compound.

Displacement pump. A type of pump in which water is induced to flow from the source of supply through an inlet pipe and inlet valve and into the pump chamber by a vacuum.

Disposal. The discharge, deposit, injection, dumping, spilling, leaking, or placing of any solid waste or hazardous waste into or on any land or water.

Disproportionation. A type of decomposition reaction in which one compound is simultaneously oxidized and reduced.

Dissociation. A chemical process that causes a molecule to split into simpler groups of atoms, single atoms, or ions. For example, water (H_2O) breaks down spontaneously into H^+ and OH^- ions.

Dissociation constant. A constant whose value depends on the equilibrium between the undissociated and dissociated forms of a molecule. In an aqueous solution, an acid (HA) will dissociate into the carboxylate anion (A^-) and hydrogen ion (H^+). $HA_{(aq)} \leftrightarrow H^+ + A^-$. The dissociation constant is usually expressed as $pK_a = -\log_{10}K_a$.

Dissolution. The process by which soluble products dissolve in a liquid.

Dissolved air flotation. Physical treatment technique whereby air is dissolved under high pressure. As the pressure drops, waste accumulates at the air/water surface and the waste is skimmed off.

Dissolved oxygen (DO). A measure of the amount of oxygen available for biochemical activity in a given amount of water. Adequate levels of DO are needed to support aquatic life. Low dissolved oxygen concentrations can result from inadequate waste treatment.

Dissolved product. For petroleum products, the water-soluble fuel components; namely, benzene, ethylbenzene toluene, and xylene.

Distillation. The separation of the components of a mixture (usually of liquids or solutions) by heating to the boiling point and condensing the resulting vapor. In multi-component mixtures such as petroleum, the various fractions have different boiling points; thus, it is possible to effect separation by condensing the vapor of each component in turn, with the lower-boiling fractions distilling off first (fractional distillation).

Distillation column. A still for fractional distillation.

Distilled water. Water formed by the condensation of steam or water vapor.

Distribution coefficient (K_d). The distribution coefficient is a variable in the retardation equation and is generally expressed as the mass of a solute on the solid phase per unit mass of solid phase divided by the concentration of the solute in solution.

Distribution free. A statistic that does not depend on which specific distribution function the observation follows; synonymous for nonparametric.

Disulfide. A compound with two sulfur atoms bonded to a radical or element.

Divalent carbon. A charged or uncharged carbon atom that has formed two covalent bonds.

Divalent metal. A metal whose atoms can chemically combine with two hydrogen atoms.

Dolomite. A carbonate sedimentary rock composed predominantly of $CaMg(CO_3)_2$.

Dominant lethal assay. A mutagenesis test used to assess the ability of a chemical to produce genetic damage. Male animals are treated with a test substance acutely (single dose) or over the entire period of sperm production. These males are then mated with females, which are examined for the number of total implantations and viable fetuses.

Dose. (1) The quantity of substance applied to a unit quantity of liquid for treatment purposes. (2) The amount of a substance absorbed by an organism, usually expressed for chemicals in the form of weight of the substance (generally in milligrams or micrograms) per unit of body weight (generally in kilograms). It is the fraction of the level of exposure to a chemical that actually enters the body following absorption.

Dose response assessment. The description of the quantitative relation between the amount of exposure to a chemical and the extent of toxic injury or disease.

DOT identification numbers. Four-digit numbers, preceded by UN or NA, that are used to identify particular substances for regulation of their transportation. See DOT publications that describe the regulations.

Double-beam spectrophotometer. An instrument that uses a photoelectric circuit to measure the difference in absorption when two closely related wavelengths of light are passed through the same medium.

Double bond. The linkage between atoms in which two pairs of electrons are equally shared.

Downflow. In an ion-exchange process, the direction of the flow of the solution being processed.

Downgradient. In the direction of decreasing static head (potential).

Downgradient well. A well which has been installed hydraulically downgradient of a site and is capable of detecting contaminant migration from the site.

Drag. The resistance offered by a liquid to the settlement or deposition of a suspended particle.

Drag coefficient. The coefficient in an empirical formula used for the determination of drag. It is the dimensionless ratio of the force per unit area to the stagnation pressure.

Drain tile. Pipes of various materials, in short lengths, laid in covered trenches underground, in most cases quite loosely and with open joints, to collect and carry off excess groundwater or to dispose of wastewater in the ground.

Drainage. General term used to denote the outflow of water from a porous medium.

Drainage basin. Region from which surface water drains into a particular stream or river.

Drainage canal. A canal built and used primarily to convey water from an area where surface and soil conditions provide no natural outlet for precipitation.

Drainage well. A well installed to drain water at or near ground surface.

Drawdown. Lowering of the water level in a well due to pumping.

Drawdown curve. In well or groundwater drainage hydraulics, the profile of the piezometric surface of the water table relating drawdown to the distance from the pumping well under a given set of pumping conditions.

Dredge sediment (spoil). Material removed from the bottom of a water body by the process of dredging which must be disposed of.

Dredging. The removal of material from the bottom of water bodies using a scooping or suction machine.

Drill log. A chronological record of the soil and rock formations which were encountered in the operation of sinking a well, with either the thickness or the elevation of the top and bottom of each formation noted.

Drilling fluid. A water- or air-based fluid used in the water-well drilling operation to remove cuttings from the hole, to clean and cool the bit, to reduce friction between the drill string and the sides of the hole, and to seal the borehole.

Drive pipe. Well casing consisting of the drive shoe and raiser. The drive pipe follows the auger bit as it advances.

Drive shoe. Steel coupling or band at the bottom edge of the casing reinforced to withstand drive pressures during cable tool and drill-through casing driver method.

Dry bulb temperature. The temperature of air measured with a dry bulb thermometer in a psychrometer to measure relative humidity.

Dry bulk density (ρ_b). The ratio of the mass of dried soil to the total soil volume.

Dry density (ρ_d). The density of soil when it is completely dry.

Dry well (dry hole). A well that does not extend into the water table or saturated zone.

Duripan. Diagnostic soil term used to describe a subsurface horizon at least half cemented by silica oxide.

Dupuit assumption. Assumptions commonly used in Darcy's Law for unconfined aquifer modeling that assumes that there is no change in hydraulic head with depth, that the hydraulic gradient equals the slope of the water table, and that the direction of groundwater flow (streamlines) is horizontal.

Dupuit equation. A mathematical expression used to approximate the flow measured in an unconfined aquifer. This equation for flow in an unconfined aquifer is described by:

$$Q = -\frac{1}{2} K(h_1^2 - h_2^2 / L)w$$

where

> Q = flow rate (M/T).
> K = saturated hydraulic conductivity (M/D/L^2).
> h_1^2 = height of water table in Well 1 above an impermeable boundary (L).
> h_2^2 = height of water table in Well 2 above an impermeable boundary (L).
> L = distance along the flow path between the two points of measurement (L).
> w = width of an aquifer (L).

Dupuit-Forchheimer assumptions. In a system of gravity flow toward a shallow sink, all the flow is assumed to be horizontal, and the velocity of the water at each point is assumed to be proportional to the slope of the water table but independent of depth.

Dusts. Solid particles generated by handing, crushing, grinding, rapid impact, and detonation of organic or inorganic materials, such as rock, ore, metal, coal, wood, and grain.

Dye. Colored substance that imparts a more or less permanent color to other materials.

Dynamic leach test. Extraction test method that provides information on rates of leachability. Solid specimens are placed in distilled water, and the leaching medium is regularly replaced after specified time intervals. A diffusion coefficient can be calculated from the test data.

Dynamic null principle. A relation used in the accurate measurement of electrical potential. The potential to be measured is balanced by an equal, opposite potential so that no current is drawn from the circuit at which the potential is being measured.

Dynamic viscosity. A measure of a fluid's resistance to tangential or shear stress.

Dyne. The absolute gram/centimeter/second unit of force defined as that force which will impart to a free mass of 1 gram an acceleration of 1 centimeter per second.

E

Eductor. A device for mixing air with water.

Effective grain size. Grain size of a theoretical body of homogeneous material of one grain size that would transmit water at the same rate as the material under consideration. It is also defined as the diameter of a grain such that 10% of the material (by weight) consists of smaller grains and 90% of larger grains.

Effective porosity (n_e). A ratio that describes the amount of interconnected pore space available for fluid transmission. It is expressed as a percentage of the total volume occupied by the interconnecting interstices. The term is sometimes analogously used to describe specific yield. The laboratory determination of this parameter is determined by saturating a soil sample, weighing it, allowing it to drain, and then re-weighing the sample. The equation for obtain effective porosity is

$$n_e = [(W_s - W_r)/(W_s - W_o)][V_v/V]100\%$$

where

n_e = effective porosity.
W_s = weight of the saturated soil sample.
W_r = weight of the soil sample after gravity drainage.
W_o = weight of the air dried soil sample.
V_v = volume of the void space.
V = total volume of the soil.

Effective rainfall. Rain that produces surface runoff.

Effective solubility. The theoretical dissolved phase concentration of a constituent that is in chemical equilibrium with other organic compounds in the matrix.

Effective storage. Volume of water available for a designated purpose.

Effective velocity. The actual or field velocity of groundwater percolating through water-bearing material. It is measured by dividing the volume of groundwater passing through a unit cross-sectional area by the effective porosity.

Effluent. (1) Solid, liquid, or gaseous wastes that enter the environment as byproducts of human-oriented processes; (2) discharge of overflow of fluid from ground or subsurface storage.

Effluent weir. A weir at the outflow end of a sedimentation basin or other hydraulic structure.

Efflux velocity. The velocity at which a gas leaves a stack. It is equal to the volume of gas issuing from the stack per second divided by the cross-sectional area of the stack.

Ejector. A device for moving a fluid or solid by entraining it in a high-velocity stream of air or water jet.

Elasticity. The extent to which a material returns to its original form or shape after being stretched, bent, strained, or otherwise deformed.

Elastomer. Term used to describe elastic polymers with rubber-like behavior. A substance that can be stretched at room temperature to at least twice its original length and, after having been stretched, returns with force to its approximate original length in a short time.

Electric double layer (double layer). A surface-layer phenomenon found at a solid/liquid interface. It is made up of the ions of one charge type which are fixed to the surface of the solid and an equal number of mobile ions of the opposite charge which are distributed through the neighboring region of the liquid.

Electric steel. Specialty high-alloy steel made by heating the components in an electric furnace; it usually contains substantial percentages of nickel, chromium, and molybdenum.

Electric well log. A record obtained from a moving electrode in a well investigation in rock; it is in the form of curves that represent the apparent values of the electric potential and electric resistivity or impedance of the rocks and their contained fluids throughout the uncased portions of a well.

Electrical conductivity. (1) The reciprocal of the resistance in ohms measured between opposite faces of a centimeter cube of an aqueous solution at a specified temperature. It is expressed as micro-ohms per centimeter at temperature degrees Celsius. (2) A surficial geophysical method whereby known current is applied to spaced electrodes in the ground and the resulting electrical resistance is used to detect changes in earth materials between and below the electrodes. Electrical conductivity is useful at sites characterized by settings having minimal quantities of high-resistance materials.

Electrical resistivity. The property of a substance to impede the flow of an electrical current through it; measured per unit length through a unit cross-sectional area.

Electrochemical cell. The combination of two electrodes arranged so that an overall oxidation-reduction reaction produces an electromotive force. Examples include dry cells, wet cells, standard cells, fuel cells, solid-electrolyte cells, and reserve cells.

Electrochemical emf. The electrical force generated by means of chemical action.

Electrochemical potential. The difference in potential that occurs when two dissimilar electrodes are connected through an external conducting circuit and the two electrodes are placed in a conducting solution so that electrochemical reactions occur.

Electrochemical reduction cell. The cathode component of an electrochemical cell.

Electrochemical series (electromotive series). A series in which the metals and other substances are listed in the order of their chemical reactivity or electrode potentials.

Electrochemistry. The study of relationships between electrical and chemical phenomena.

Electrochromatography. A type of chromatography that utilizes the application of an electric potential to produce an electric differential.

Electrocratic. A term used in colloid chemistry to denote a dispersion of insoluble solid particles in a liquid whose stability is maintained by either positive or negative electric charges on the particles.

Electrode. A material used in an electrolytic cell to enable the current to enter or leave the solution. Each cell has a positive electrode (anode) and a negative electrode (cathode).

Electrode potential. The voltage existing between an electrode and the solution or electrolyte in which it is immersed.

Electrolysis. The chemical splitting of an electrolyte by passing an electric current through the solution.

Electrolyte. A chemical that dissociates into positive and negative ions when dissolved in a solution, thereby increasing the electrical conductivity of the solution.

Electrolytic conductivity. The conductivity of a medium in which the transport of electric charges, under electric potential differences, occurs by particles of atomic or larger size.

Electrolytic potential. The difference in potential between an electrode and the immediately adjacent electrolyte.

Electrometric titration. A titration in which the end point is determined by observing the change of potential of an electrode immersed in the solution titrated.

Electromotive force (emf). The difference in electric potential that exists between two dissimilar electrodes immersed in the same electrolyte.

Electron acceptor. In biochemical oxidation, the electron acceptor is the oxidizing material to which electrons are transferred from an electron donor.

Electron capture. The process in which an atom or ion passing through a medium loses or gains one or more orbital electrons.

Electron capture detector. A detector in which a voltage is produced through ionization of a special carrier gas, usually nitrogen or argon/methane; organic compounds (e.g., PCBs, alkyl halides, carbonyls, nitriles, nitrates, and organometals) are detected through their capacity to absorb electrons and therefore impede electrical current. The resulting reduction in electrical current is monitored by an electrometer and is displayed as a positive signal on a meter or strip chart recorder. The detector is virtually insensitive to aliphatic hydrocarbons, alcohols, and ketones.

Electron configuration. The orbital and spin arrangement of the electrons of an atom.

Electron donor. In biochemical oxidation, electrons that are transferred from a reduced substance (electron donor) to an oxidizing material (electron acceptor).

Electron mass (m_e). A fundamental physical constant equal to 9.109×10^{-31} kg.

Electron number. The number of electrons in an ion or atom.

Electron pair. A pair of valence electrons which form a nonpolar bond between two neighboring atoms.

Electron spectroscopy. The study of the energy spectra of photoelectrons or Auger electrons emitted from a substance upon bombardment by electromagnetic radiation, electrons, or ions.

Electronegative potential. The potential of an electrode with respect to the hydrogen electrode.

Electronic band spectrum. The bands of spectral lines associated with a change of electronic state of a molecule.

Electronic magnetic moment. The total magnetic dipole moment associated with the orbital motion of all the electrons of an atom and the electron spins.

Electronic spectrum. The spectrum resulting from emission or absorption of electromagnetic radiation during changes in the electron configuration of atoms, ions, or molecules.

Electroplating. A type of electrodeposition in which a metal or plastic is coated with a film of metal for corrosion protection or for decorative purposes.

Element. A unique arrangement of fundamental units of matter having characteristic properties; 106 elements are presently known, of which 92 occur in nature, the others being synthetic. The smallest amount of an element that can exist is the atom.

Elutriation. The process of separating particles using specific gravity differences between particles when suspended in a fluid.

Eluviation. The movement of material out of a portion of a soil profile.

Emission flame photometry. A form of flame photometry in which a sample solution to be analyzed is aspirated into a hydrogen-oxygen or acetylene-oxygen flame. The line emission spectrum is formed, the line or band of interest is isolated with a monochromator, and its intensity is measured photoelectrically.

Emission lines. The spectral lines resulting from emission of electromagnetic radiation by atoms, ions, or molecules during changes from excited states to states of lower energy.

Emission spectrometer. A spectrometer that measures the percent concentration of pre-selected elements in samples of metals and other materials.

Emission spectrum. The electromagnetic spectrum produced when radiations from any emitting source, excited by any of various forms of energy, are dispersed.

Empirical. Relying upon or gained from experiment or observation.

Emulsification. The process of dispersing one liquid in a second immiscible liquid.

Emulsifier. A surface-active substance that stabilizes (reduces the tendency to separate) a suspension of droplets of one liquid in another liquid which otherwise would not mix with the first. Some or all applications may be classified by the EPA as Restricted Use Pesticides.

Emulsifying agent. An agent capable of modifying the surface tension of emulsion droplets to prevent coalescence.

Emulsion. A heterogeneous liquid mixture of two or more liquids not normally dissolved in one another but held in a colloidal suspension by forceful agitation or by emulsifiers.

End point. That point in a chemical titration when no further addition of titration is necessary; it is usually denoted by a change in color of an added substance called an indicator.

Endosulfan. An insecticide (acaricide) which controls aphids, bollworms, bugs, whiteflies, leafhoppers, and slugs on deciduous and citrus trees, small fruits, vegetables, forage crops, nut crops, oil crops, fiber crops, grains, tobacco, coffee, tea plants, forests, and ornamentals.

Endothermic. The term used to characterize a chemical reaction which requires absorption of heat from an external source.

Enhanced spectral line. A spectral line of a very hot source, the intensity of which is much greater than that of a line in a flame or arc spectrum.

Entrained. Particulates or vapor transported along with flowing gas or liquid.

Entrance slit. A narrow slit through which light passes when entering a spectrometer.

Environmental priorities initiative (EPI). An initiative whereby facility assessment and site investigations at RCRA are done under the Superfund program.

Enzyme. A catalyst produced by a living organism.

Enzyme immunoassay. An assay procedure based on the reversible and noncovalent binding of an antigen by a specific antibody, in which one of the reactants is labeled with an enzyme.

Enzyme-linked. An enzyme immunoassay in which one of the reactants is absorbed onto the surface of the wells of a microtiter plate.

Eocene. A geologic epoch between 57.8 and 36.6 million years ago.

EPA hazardous waste number. A number assigned by EPA to waste that is hazardous by definition; to each hazardous waste listed in 40 CFR 261 Subpart D from specific and nonspecific sources identified by the EPA (F, K, P, U); and to each characteristic waste identified in 40 CFR 261 Subpart C, including wastes with ignitables (D001), reactive (D002), corrosive (D003), and EP toxic (D004–D017) characteristics.

EPA identification number. A number assigned by the EPA to each generator, transporter, and treatment, storage, or disposal facility. The numbers are facility-specific, except for the transporter, who has one number for all his operations.

Ephemeral stream. A stream that flows only in direct response to precipitation.

Epidemiology. Study of the distribution and causes of diseases and injuries in human populations.

Epipedon. The uppermost soil horizon.

Epithermal ore deposits. Genetic classification of an ore that is of hydrothermal origin and forms at shallow depths (<300 ft.) and temperatures of 50 to 200°C. Examples are native gold, sylvanite, cinnabar, and stibnite.

Epoxidation. A reaction in which an epoxy compound is created.

Epoxy resin. A versatile synthetic resin made by reacting an epoxide compound (usually epichlorohydrin) with a hydroxyl-containing substance such as bisphenol A or a polyhydric alcohol (glycerol).

Equation of state. A mathematical expression which defines the physical state of a homogeneous substance by relating volume to pressure and absolute temperature for a given mass of the material.

Equilibrium diagram. A phase diagram of the equilibrium relationship between temperature, pressure, and composition of a system.

Equilibrium leach test. An extraction test method using batch extraction. The leaching medium is distilled water. A ground solid sample is mixed at a 4:1 liquid-to-solid ratio and agitated for 7 days.

Equilibrium potential. The point at which forward and reverse reaction rates are equal in an electrolytic solution.

Equipment blank. A chemically pure solvent (typically reagent grade water) which is passed through field sampling equipment that is in contact with the sample during collection. The equipment blank is returned to the laboratory for analysis to determine the effectiveness of equipment decontamination procedures.

Equipotential (equal pressure) lines. Lines drawn between points of equal pressure.

Equivalence point. The point in a titration at which the amounts of titrant and material being titrated are chemically equivalent.

Equivalent weight. A term used in chemistry to denote the weight of a compound that contains one gram atom of available hydrogen or its chemical equivalent. A compounds' equivalent weight is equal to its molecular weight (MW) divided by an integer (Z) which is dependent on the compound. For acids, this integer is equal to the number of moles of H^+ obtained from one mole of the acid; for bases, it is equal to the number of moles of H^+ that will react with one mole of the base.

Error. The difference between an experimental and the "true" value.

Escarpment. A more or less continuous line of cliffs or steep slopes facing in one general direction and caused by erosion or faulting.

Esker. A long, winding, gravelly ridge deposited by a stream flowing in a channel in the ice of a glacier or below a glacier.

Essential elements. Elements that are necessary for the growth of microorganisms and other biota. Essential elements include nitrogen, phosphorous, potassium, calcium, magnesium, sulfur, iron, manganese, copper, zinc, boron, molybdenum, chlorine, and cobalt.

Ester. A compound formed by replacement of the hydrogen of an acid by an aklyl or other hydrocarbon group, usually derived from a hydroxyl-bearing compound.

Esterification. The process of making an ester from an alcohol.

Esters. A class of organic compounds derived by the reaction of an organic acid with an alcohol.

Estuary. The mouth of a stream which serves as a mixing zone for fresh and ocean water. Mouths of streams which are temporality separated from the ocean by sandbars are considered as estuaries by the State Water Resources Control Board. Estuarine waters are considered to extend seaward, and significant mixing of fresh water and seawater occurs in the open coastal waters.

Etch. A removal of specific material (such as portions of a given layer) through a chemical reaction.

1,2-Ethanedithiol ($HSCH_2CH_2SH$). A liquid, freely soluble in alcohol and in alkalines, which is used as a metal complexing agent.

Ethanol (C_2H_5OH). A colorless liquid, miscible with water, which is used as a reagent and solvent.

Ethanolamine ($NH_2(CH_2)_2OH$). A colorless liquid, miscible in water; used in scrubbing H_2S and Co_2 from petroleum gas streams, for dry cleaning, in paints, and in pharmaceuticals.

Ether. A class of organic compounds characterized by the presence of an oxygen atom covalently bound between two carbon atoms.

Etherification. The process by which ether is made from an alcohol.

Ethers. Organic compounds formed by the treatment of alcohols with strong dehydrating agents.

Ethoxide (ethylate). A compound formed from ethanol by replacing the hydrogen of the hydroxy group by a monovalent metal.

Ethyl acetate ($CH_3COOC_2H_5$). A colorless liquid, slightly soluble in water, which is used as a medicine, reagent, and solvent; also known as acetic ester, acetic ether, and acetidin.

Ethyl acrylate ($C_5H_8O_2$). A colorless liquid used to manufacture chemicals and resins.

Ethyl borate ($B(OC_2H_5)$). A salt of ethanol and boric acid used in antiseptics, disinfectants, and fireproofing.

Ethyl bromide (C_2H_5Br). A colorless liquid used as a refrigerant and in organic synthesis.

Ethyl butyrate ($C_3H_7COOH_2H_5$; butyric ether). A colorless liquid used in flavoring extracts and perfumery.

Ethyl caprate ($C_5H_{11}COOC_2H_5$). A colorless liquid used in the manufacture of wine and cognac.

Ethyl enanthate ($CH_3(CH_2)_5COOC_2H_5$). A clear oil that is soluble in alcohol, chloroform, and ether. It is used as a flavor for liquors and soft drinks.

2-Ethyl hexoic acid ($C_4H_9CH(C_2H_5)CH_2Br$). A liquid that is slightly soluble in water and is used as an intermediate to make metallic salts for paint and varnish driers, esters for plasticizers, and light metal salts for conversion of some oils to grease.

Ethyl mercaptan (C_2H_5SH). A colorless liquid with a boiling point of 36°C; also known as ethanethiol, ethyl sulfhydrate, and thioethyl alcohol.

Ethyl oxalate ($(COOC_2H_5)_2$). An oily, unstable liquid that is combustible, miscible with organic solvents, and slightly soluble in water. It is used as a solvent for cellulosic and resins and as an intermediate for dyes and pharmaceuticals.

Ethyl sulfide ($(C_2H_5)_2S$). An oily liquid used as a solvent and in organic synthesis.

Ethylbutene ($CH_3CH_2(C_2H_5)CCH_2$). A stable, colorless liquid, miscible in most organic solvents and insoluble in water, which is used in organic synthesis.

Ethylbutyl acetate ($C_2H_5CH(C_2H_5)CH_2O_2CCH_3$). A colorless liquid used as a solvent for resins, lacquers, and nitrocellulose.

Ethylbutyl alcohol ($(C_2H_5)_2CHCH_2OH$). A stable, colorless liquid, miscible in most organic solvents and slightly water soluble, which is used as a solvent for resins, waxes, and dyes and in the synthesis of perfumes, drugs, and flavorings.

Ethylbutyl ketone (C₂H₅COC₄H₉; 3-heptanone). A colorless liquid used in solvent mixtures.

Ethylene (C₂H₄). A colorless, flammable gas used as an agricultural chemical and for the manufacture of organic chemicals and polyethylene.

Ethylene dibromide (BrCH₂CH₂Br). A colorless, poisonous liquid which is soluble in water. It is used in medicine, as a solvent in organic synthesis, in anti-knock gasoline, and as a fumigant.

Ethylene dichloride. A saturated chlorinated hydrocarbon liquid made by catalytic reaction of ethylene with chlorine.

Ethylene glycol *bis* (C₄H₄Cl₆O₄; trichloroacetate). A white solid used as an herbicide for cotton and soybeans.

Ethylene glycol diacetate (CH₃COOCH₂CH₂OOCCH₃). A liquid used as a solvent for oils, cellulose esters, and explosives.

Ethylene propylene diene monomer (EPDM). A synthetic elastomer based on ethylene, propylene, and a small amount of nonconjugated diene that provides sites for vulcanization.

Ethylene oxide ((CH₂)₂O). A colorless gas, soluble in organic solvents and miscible in water, used in organic synthesis, for sterilizing, and for fumigating.

Ethylene resin. A thermoplastic material composed of polymers of ethylene.

Ethyleneimine (C₂H₄NH). A highly corrosive liquid, used as an intermediate in fuel oil production, refining lubricants, textiles, and pharmaceuticals.

2-Ethylhexyl bromide (C₄H₉CH(C₂H₅)CH₂Br). A water-white, water-insoluble liquid.

2-Ethylhexyl chloride (C₄H₉CH(C₂H₅)CH₂Cl). A colorless liquid used to synthesize cellulose derivatives, pharmaceutical, resins, insecticides, and dyestuffs.

Eucaryotes. An organism having one or more cells with well-defined nuclei.

Evaporation. The process by which a liquid enters the vapor (gas) phase.

Evaporation rate. The quantity of water, expressed in terms of depth of liquid water, evaporated from a given water surface per unit of time.

Evaporite. A sedimentary rock formed due to the evaporation of sea water.

Evapotranspiration. The loss of water from a land area through transpiration of plants and evaporation from the soil.

Exchange adsorption. A specific type of adsorption that is characterized by the electrical attraction between the adsorbate and the surface of a material.

Exchange narrowing. The phenomenon in which a spectral line is split and thereby broadened by some variable perturbation. The broadening may be narrowed by a dynamic process that exchanges different values of the perturbation.

Exchange reaction. A reaction in which two atoms or ions exchange places either in two different molecules or in the same molecule.

Exchangeable acidity. That portion of the total acidity that can be removed from soil cation exchange sites by soil equilibrium with a neutral unbuffered salt such as potassium chloride or sodium chloride.

Excitation. The process in which an atom or molecule gains energy from electromagnetic radiation or by collision, raising it to an excited state.

Exhaust ventilation. The removal of air, usually by mechanical means, from any space. The flow of air between two points is due to the occurrence of a pressure difference between the two points. This pressure difference will cause air to flow from the high-pressure to the low-pressure zone.

Exhaustion point. In an ion-exchange process, the state of an adsorbent when it can no longer produce a useful ion exchange.

Existing portion. The land surface of an existing waste management unit, included in the original Part A permit application, on which wastes have been placed prior to the issuance of a permit.

Existing TSD facility. A treatment, storage, or disposal (TSD) facility that began operation or construction on or before November 19, 1980. An existing facility may qualify for interim status.

Exocyclic double bond. A double bond that is connected to and external to a ring structure.

Expansion coupling. A pipe coupling which permits relative movement of the two joined pipes.

Explosivity meter (combustible gas indicator). An instrument that detects gas vapor in air and indicates whether the test atmosphere contains a flammable level of gas vapor.

Exposure. Level of exposure to a substance in the vicinity of a portal of entry to the body (e.g., lungs, mouth, skin) that may be available for absorption.

Exposure assessment. A description of the nature and size of various populations' exposure to a chemical and the magnitude and duration of the exposure.

Exposure level. The level or concentration of a physical or chemical hazard to which a person is exposed.

Exposure limits. Concentration of substances (and conditions) under which it is believed that nearly all workers may be repeatedly exposed day after day without adverse effects.

Ex situ. Moved from its original place; excavated; removed or recovered from the subsurface.

Ex situ **biodegradation**. The microbiological degradation of volatile organic compound in soil that is accomplished on the ground surface.

Extender. Material used to dilute or extend or change the properties of resins, ceramics, paints, and rubber.

External electron transfer. The transfer of electrons to and from some agent other than from the halogenated compound.

Extractant. The liquid used to remove a solute from another liquid.

Extraction. A physical treatment process where dissolved or absorbed substances are transferred from a liquid or solid phase of a solvent.

Extraction procedure (EP) toxicity. One of the characteristics (along with ignitability, reactivity, and corrosivity) that make a waste hazardous. The EP toxic list includes maximum concentrations for 14 constituents which, if exceeded, would make a waste hazardous. Effective September 1990, EP toxicity was replaced by the toxicity characteristic.

Extraction well. A well employed to extract fluids (water, gas, free product, or a combination of these) from the subsurface.

Extrapolation. The process of predicting the response of an organism to some conditions of exposure to a substance based on the magnitude of the response seen at other conditions of exposure or in another species.

F

Fabric (rock). The spatial and geometric configuration of all those components that make up a deformed rock. It is an all-encompassing term that describes the shapes and characters of individual parts of a rock mass and the manner in which these parts are distributed and oriented in space.

Face velocity. The average air velocity into an exhaust system measured at the opening into the hood or booth.

Facility. All contiguous land, structures, or other appurtenances and improvements on the land used for treating, storing, or disposing of hazardous wastes. A facility may consist of several treatment, storage, or disposal operational units (e.g., one or more landfills, surface impoundments, or combinations of them).

Facultative. Used to describe organisms that are able to grow in either the presence or absence of a specific environmental factor (e.g., oxygen).

Facultative anaerobes. Microorganisms that can grow in either the presence or absence of molecular oxygen. In the absence of oxygen, these microorganisms can utilize another compound (e.g., sulfate or nitrate) as a terminal electron acceptor.

Fahrenheit. A temperature scale in which 32° marks the freezing point and 212° the boiling point of water at a 760-mm barometric pressure. To convert to centigrade (Celsius), subtract 32 and multiply by 5/9.

False body. The property of certain colloidal substances, such as paints and printing inks, of solidifying when left standing.

False negative. A statistical term used to describe an event, such as contamination, that has occurred but has not been detected.

False positive. A statistical term used to describe an event, such as contamination, where the statistical analysis indicates that contamination has occurred when, in fact, it has not occurred.

Fast neutron spectrometry. Neutron spectrometry in which nuclear reactions are produced by or yield fast neutrons.

Fatty acid (C_nH_{2n} + COOCH). An aliphatic acid, either saturated or unsaturated, whose molecule consists of an alkyl chain containing from 1 to over 30 carbon atoms, terminating in a carboxyl group (COOH).

Fatty alcohol. A high-molecular-weight, straight-chain, primary alcohol derived from natural fats and oils.

Feldspar. A clay-like material containing potassium as well as aluminum and silica.

Fermentation. The decomposition of organic substances by microorganisms and/or enzymes.

Ferrallitization (desilication, ferritization, allitization). The chemical migration of silica from a soil horizon.

Ferrate. A multiple iron oxide that is combined with another oxide.

Ferric. The term for a compound of trivalent iron.

Ferric ammonium oxalate ($(NH_4)_3Fe(C_2O_4)_3 \cdot 3H_2O$). A green, crystalline material, soluble in water and alcohol, which is sensitive to light. It is used in blueprint photography.

Ferric arsenate ($FeAsO_4 \cdot 12H_2O$). A green or brown powder, insoluble in water and soluble in dilute material acids, which is used as an insecticide.

Ferric chloride ($FeCl_3$). Brown crystals which are soluble in water, alcohol, and glycerol; used as a coagulant for sewage and industrial wastes, as an oxidizing and chlorinating agent, as a disinfectant, in copper etching, and as a mordant.

Ferric nitrate ($Fe(NO_3)_3 \cdot 9H_2O$). Colorless crystals that are soluble in water and used as a dyeing mordant, in tanning, and in analytical chemistry.

Ferric phosphate ($FePO_4 \cdot 2H_2O$). Yellow crystals, insoluble in water and soluble in acid; used in medicines and fertilizers.

Ferric resinate. A reddish-brown, water-insoluble powder used as a drier for paints and varnishes.

Ferric stearate ($Fe(C_{18}H_{35}O_2)_3$). A light-brown, water-insoluble powder used as a varnish drier.

Ferroalloy. Alloy containing varying proportions of iron and some other metal, such as molybdenum, manganese, titanium, etc.; commonly added to steel for purposes of strength and heat resistance.

Ferrous. Term or prefix used to denote compounds of iron in which iron is in the divalent (+2) state.

Ferrous hydroxide ($Fe(OH)_2$). A water-insoluble solid that turns reddish-brown as it oxidizes to ferric hydroxide.

Fiber. Elongated crystalline structure ranging in length from a few millimeters to several feet and in diameter from 1 micron to 0.05 inch.

Fibric soil (peat). A soil in which organic fibers comprise two thirds of the mass. Peat yields a clear solution when extracted with sodium pyrophosphate.

Fick's First Law. An equation describing the rate at which a gas transfers into solution. The change in concentration of gas in solution is proportional to the product of an overall mass transfer coefficient and the concentration gradient. Fick's First Law is expressed as:

$$F = -D(dC/dx)$$

where

F = mass flux or the mass of the solute per unit area per unit time ($M/L^2/T$).

D = diffusion coefficient (L^2/T).

dC/dx = the concentration gradient which is negative in the direction of diffusion.

Fick's Second Law. An equation relating the change of concentration with time due to diffusion to the change in concentration gradient with distance from the source of concentration.

Field air capacity. The fractional volume of air in a soil at the field capacity water content.

Field bank. A laboratory-prepared sample of Type II, reagent-grade water or pure solvent which is transported to the sampling site for use in quality assurance/ quality control evaluation of field sampling procedures.

Field capacity (field moisture capacity). The presumed water content remaining in soil 2 or 3 days after a soil has been saturated. This percentage is expressed on a weight (gravimetric) or volume basis.

Filter cloth. A fabric stretched around the drum of a vacuum filter.

Filter pack. Sand that is placed in the annulus of the wall between the borehole wall and the well screen to prevent formation material from entering the well screen. The filter pack should extend some distance above the well screen.

Filter press. A press operated mechanically for partially separating water from solid materials.

Filter spectrophotometer. A spectrophotographic analyzer of spectral radiations in which a filter is used to isolate narrow portions of the spectrum.

Filter structure. The splitting of spectral lines in atomic and molecular spectra caused by the spin angular momentum of the electrons and the coupling of the spin to the orbital angular momentum.

Filtrate. The liquid which has passed through a filter.

Filtration. (1) A physical treatment process by which suspended solids are removed from a fluid by passage through porous media with a force of gravity; (2) the process of passing a liquid through a filtering medium.

Fine-grained soils. Soils in which more than half of the material passes through a number 200 sieve.

Fine sand. Sediment particles having diameters between 0.125 and 0.250 millimeters.

Fischer-Tropsh process. A process for the manufacture of liquid hydrocarbon fuels that consists of a catalyzed reaction of carbon monoxide and hydrogen which yields aliphatic hydrocarbons and oxygen-containing organics. The gaseous mixture of hydrogen and carbon monoxide can be obtained from petroleum or coal (water gas or synthesis gas). A number of metallic catalysts have been successfully used. It is now being applied to certain developments in coal gasification.

Flame emission spectroscopy. A flame photometry technique in which the solution containing the sample to be analyzed is optically excited in an oxyhydrogen or oxyacetylene flame.

Flame excitation. The use of a high-temperature flame to excite spectra emission lines from alkali and alkaline-earth elements and metals.

Flame ionization detector. A detector in which molecules are ionized in a hydrogen flame. The ions and electrons formed decrease electrical resistance in a gap between two electrodes and permit the flow of current, which is amplified and displayed on a meter or strip chart recorder.

Flame spectrometry. A procedure used to measure the spectra or to determine wavelengths emitted by flame-excited substances.

Flame spectrophotometry. A method used to determine the radiation intensity of various wavelengths in a spectrum emitted by a chemical inserted into a flame.

Flame spectrum. An emission spectrum obtained by evaporating substances in a nonluminous flame.

Flammable. A material with a flash point less than 100°F (and a vapor pressure of not over 50 psi at 100°F).

Flammable liquid. A liquid with a flash point below 100°F (37.8°C), excluding gases.

Flash back. Phenomenon characterized by vapor ignition and flame travel back to the vapor source (the flammable liquid).

Flash point. Lowest temperature at which vapors from a volatile liquid will ignite upon the application of a small flame under specified conditions.

Flash set. Premature stiffening of a cement slurry.

Flash spectroscopy. The study of the electronic states of molecules after they absorb energy from an intense, brief light flash.

Floaters. Light-phase organic liquids in groundwater capable of forming an immiscible layer that can "float" on the water table.

Floating pan. An evaporation pan floating in a body of water.

Floc. Small gelatinous masses formed in a liquid by the reaction of a coagulant added through biochemical processes, or by agglomeration.

Flocculation. A physical treatment process in which small suspended particles are transformed into larger, more settled ones by the addition of a chemical, typically alum, lime, or a polyelectrolyte.

Flocculation agent. A coagulating substance which, when added to water, forms a flocculant precipitate which will entrain suspended matter and expedite sedimentation.

Floodplain. Land adjacent to a river that is covered by water when a river overflows its banks during flooding.

Flotation agent. A chemical which alters the surface tension of water or makes it froth easily.

Flow cell. A sensing element or combination of elements, such as electrodes, immersed in a flowing fluid and which continuously measures some property of the fluid such as pH, dissolved oxygen, or electrical conductivity.

Flow net. A plot of intersecting equipotential lines and flow lines representing a two-dimensional, steady flow through porous media.

Flow path. Direction in which groundwater is moving.

Flow tube. A calibrated flow-measuring device made for a specific range of flow velocities and fluids.

Flow-weighted sampling. Samples taken in a manner that allows determination of mass emissions (i.e., samples taken in proportion to the rate of flow of a river or stream).

Fluid. Any material or substance that changes shape or direction uniformly in response to an external force imposed upon it. The term applies not only to liquids but also to gases and to finely divided solids.

Fluid mechanics. The science of the motion of fluids, based on physical analysis and experimental verification.

Fluid potential. Mechanical energy per unit mass of fluid at any given point in space and time with respect to an arbitrary state and datum. A loss of fluid potential occurs as fluid moves from a region of high potential to one of low potential.

Fluid pressure (ρ). The force per unit area on a point; for a water column, it is the force per unit area that acts at that point on the water column.

Fluidized bed. Combustion system in which fuel of low calorific value is combusted by blowing air through the mixture of fuel and inert fire bed to which it is introduced.

Fluoranthene. One of the 126 priority pollutants listed by the EPA under Section 307(a) of the Clean Water Act.

Fluorene ($C_{13}H_{10}$). A hydrocarbon chemical present in the middle oil fraction of coal tar.

Fluorescence. The ability of a material to absorb radiant energy of one wavelength and emit the energy as radiation at a longer wavelength.

Fluorescence spectra. The emission spectra of fluorescence in which an atom or molecule is excited by absorbing light and then emits light of characteristic frequencies.

Fluoroacetate. An acetate in which carbon-connected hydrogen atoms are replaced by fluorine atoms.

Fluorocarbon. Any of a broad group of organic compounds analogous to hydrocarbons, in which all or most of the hydrogen atoms of a hydrocarbon have been replaced by fluorine.

Fluorocarbon resin. A polymeric material made up of carbon and fluorine with or without other halogens or hydrogen.

Fluorochemical. Any chemical compound containing fluorine.

Fluosilicate. A salt derived from fluorosilic acid.

Flushing. The process by which contaminant concentrations in a body of water are diluted by river inflow and, where applicable, tidal exchange of "new" uncontaminated water.

Flute. To remove sorbed materials (chemicals) from a sorbed (column) by means of a carrier gas. A compound is said to elute when it emerges from the outlet end of the chromatographic column into the detector.

Fluvial deposits. Sediments deposited by physical processes in river channels or floodplains. This term is synonymous with alluvial deposits.

Fluvial erosion. Erosion caused by the action of streams.

Flux. Rate of transfer of a quantity (water, heat, etc.) across a surface.

Foam. A liquid or solid emulsion-like system in which gas is more or less uniformly distributed. The entrapped vacuoles range in size from colloidal to optically visible.

Foaming adjuvant. A surface-active substance that produces a fast-draining foam to provide maximum contact of the spray with a plant surface, to insulate the surface, and to reduce the rate of evaporation. Used to enhance herbicide action and to reduce drift of sprays.

Foot-pound. The amount of energy required to lift 1 pound a vertical distance of 1 foot.

Forensic chemistry. The application of chemistry to the study of materials or problems in cases where the findings may be presented as technical evidence in a court of law.

Formaldehyde sodium bisulfite (CH_3NaO_4S). A compound used as a fixing agent for fibers containing keratin, in metallurgy for flotation of lead-zinc ores, and in photography.

Formation. An assemblage of rock masses grouped together into a unit that is convenient for description or mapping.

Formic acid (HCOOH). A colorless, toxic, corrosive liquid soluble in water, ether, and alcohol.

Formula. An expression, usually in the form of an algebraic equation, of the value of a given variable in terms of one or more independent variables and the necessary constants. A method of reasoning stated in the form of an equation.

Formula weight. The gram-molecular weight of a substance.

Four-degree calorie. The heat needed to change the temperature of 1 gram of water from 3.5 to 4.5°C.

Fourier's Law. A law that states that the rate of heat conduction is proportional to the temperature gradient.

Fractional air content. A measure of the relative air content of a soil; synonymous with air-filled porosity.

Fractional distillation. A method to separate a mixture of several volatile components of different boiling points. The mixture is distilled at the lowest boiling point, and the distillate is collected as one fraction until the temperature of the vapor rises, showing that the next higher boiling component of the mixture is beginning to distill. That component is then collected as a separate fraction.

Fractionation. The separation of a mixture in successive stages, with each stage removing from the mixture some proportion of one of the substances.

Fracture. A general term for any break in a rock, whether or not it causes displacement, due to mechanical failure caused by stress.

Fracture spring. A spring in which water flows from relatively large openings in rocks.

Fracture system. A group of fractures (faults, joints, or veins) consisting of one or more sets that are usually intersecting or interconnected.

Fracture zone. A thickness of strata that has undergone mechanical failure due to stress (e.g., cracks, joints, and faults).

Free available chlorine. The amount of chlorine available as dissolved gas, hypochlorous acid, or hypochlorite ion that is not combined with an organic compound.

Freeboard. The vertical distance between the top of a tank or surface impoundment dike and the surface of the liquid waste it contains.

Free chlorine residuals. A reference to the chlorine, hypochlorous acid, and hypochlorite ion in a solution.

Free cyanide. A cyanide not combined as part of an ionic complex.

Free iron. The percentage of total iron in a soil that occurs as hydrous oxides.

Free liquids. Liquids which readily separate from the solid portion of a waste under ambient temperature and pressure.

Free product. Fuel product accumulated on top of the groundwater and which is recoverable by well withdrawal methods; free product is often mobile.

Free radical. A molecular ion that possesses one or more unpaired electrons.

Free surface. The boundary surface of a liquid in contact with the atmosphere.

Freeze. To solidify a liquid by removal of heat.

Freezing point. The temperature at which a liquid and a solid is in equilibrium.

Freezing point depression. The lowering of the freezing point of a solution compared to the pure solvent.

French drain. An underground passageway for water through the interstices among stones placed loosely in a trench.

Frequency histogram. A graphical display that shows the distribution of a set of data. The height of each bar is proportional to the number of data values that fall within the interval of each bar.

Friable. In soil morphology, a technique used to describe qualitatively the moisture consistency of a soil. It is determined by placing the soil between the thumb and finger and applying pressure to the soil; if gentle thumb and finger pressure is required to crush the aggregate, it is considered friable.

Friction loss. The head lost by water flowing in a stream or conduit as the result of the disturbances set up by the contact between the moving water and its containing conduit and by intermolecular friction.

Friedel-Crafts reaction. A substitution reaction, catalyzed by aluminum chloride, in which an alkyl or an acyl group replaces a hydrogen atom of an aromatic nucleus to produce hydrocarbon or a ketone.

Frigid soil. A soil with a temperature between 0 and less than 8°C.

Fringe water. Mobile water occurring in the capillary fringe above the water table that completely fills the smaller interstices.

Fritted glass. Porous glass.

Frother. A substance used in flotation processes to make air bubbles sufficiently permanent, principally by reducing surface tension.

Fuel cell catalyst. A substance, such as platinum, silver, or nickel, from which the electrodes of a fuel cell are made.

Fuel cell electrolyte. A substance that conducts electricity between the electrodes of a fuel cell.

Fugitive water. Leakage from impounding reservoirs or an irrigation system.

Fulvene (C_6H_6). A yellow oil that is an isomer of benzene.

Fumes. Particulate matter consisting of the solid particles generated by condensation from the gaseous state, generally after violation from melted substances, and often accompanied by a chemical reaction, such as oxidation.

Fumigant. A chemical compound which acts in the gaseous state to destroy insects and their larvae and other pests.

Functional group. An atom, or group of atoms acting as a unit, that has replaced a hydrogen atom in a hydrocarbon molecule and whose presence imparts characteristic properties to this molecule.

Functionality. The ability of a compound to form covalent bonds.

Fungi. Aerobic, multicellular, nonphotosynthetic, heterotrophic microorganisms. The fungi include mushrooms, yeast, molds, and smuts. Most fungi are saprophytes, obtaining their nourishment from dead organic matter. Along with bacteria, fungi are the principal organisms responsible for the decomposition of carbon in the biosphere. Fungi have two ecological advantages over bacteria: (1) they can grow in low moisture areas, and (2) they can grow in low pH environments.

Fungicide. Class of pesticides used to kill fungi, primarily those that cause diseases of plants.

Furan. A group of organic compounds containing a ring of carbon atoms and one oxygen atom.

Furan resin. A thermosetting polymer made from furfuryl alcohol in two stages. It is first partially polymerized to a viscous liquid, to which such fillers as asbestos or wood fibers are added; it is then further polymerized to a hard solid form at the site of its use by adding an inorganic acid which acts as a catalyst.

Furnace black. A carbon black formed by the partial combustion of liquid and gaseous hydrocarbons in a closed furnace with a deficiency of oxygen.

G

Gabbro. A coarse-grained igneous rock composed primarily of feldspar and augite.

Gallery. An underground structure designed and installed to collect subsurface water.

Gallium halide. Compound formed by bonding gallium to chlorine, bromine, iodine, fluorine, or astatine.

Galvanometer. Instrument used for detecting the existence and/or measuring the strength of a small electric current by movements of a magnetic needle of a coil within a magnetic field.

Gamma flux density. Number of gamma rays emitted by a nucleus in a transition between two energy levels.

Gamma radiation. Electromagnetic radiation that travels at the speed of light. The unit of gamma radiation is the photon.

Gamma ray logging. A method of logging within a borehole that measures the amount of natural radioactivity present in a formation vs. depth. Gamma ray logging reflects the shale content of the formations, because radioactive elements (e.g., K^{40}, U^{238}, Th^{232}) tend to concentrate in shale.

Gamma rays. High-energy, short-wavelength electromagnetic radiation emitted by a nucleus. Energies of gamma rays are usually between 0.010 and 10 Mev.

Gangue. Valueless rock or mineral aggregates in an ore.

Gas. The state of matter in which the material has very low density and viscosity. Gas can expand and contract greatly in response to changes in temperature and pressure; it easily diffuses into other gases; and it readily and uniformly distributes itself throughout any container.

Gas adsorption. The concentration of a gas on the surface of a solid substance due to attractive forces between the surface and the gas molecules.

Gas analysis. Analysis of the constituents or properties of a gas (either pure or mixed). The composition can be measured by chemical adsorption, combustion, electrochemical cells, indicator papers, chromatography, and mass spectroscopy.

Gas chromatography (GC). A method of chemical analysis based on the vaporization of a liquid sample followed by the separation of the gaseous components that are then identified and quantified. It is commonly used as a quantitative analytical technique for volatile compounds.

Gas lift. The mechanical process of lifting a column of water from a well where pressurized gas is used as the lifting agent.

Gas plant. A manufacturing facility where gas for utility distribution is made from coal and/or oil using any number of a variety of processes.

Gas solid chromatography. A form of gas chromatography in which the moving phase is a gas and the stationary phase is a surface-active sorbet.

Gas solubility. The extent that a gas dissolves in a liquid to produce a homogeneous system.

Gasification. The production of gaseous fuels formed by the reaction of hot carbonaceous materials with air, steam, or oxygen.

Gasoline. A refined mixture of petroleum hydrocarbons having an octane number ranging from about 60 to 100 and a carbon number between C_5 and C_{14}.

Gate valve. A valve regulated by the position of a circular plate.

Gatterman-Koch synthesis. Synthesis in which aldehydes form when an aromatic hydrocarbon is heated in the presence of hydrogen chloride, certain metallic chloride catalysts, and either carbon monoxide or hydrogen cyanide.

Gauge. The thickness of steel used to manufacture a drum. The lower the gauge, the thicker the material; also used to measure glove thickness in inches.

Gauging. The determination of the quantity of water flowing per unit of time in a stream channel, conduit, or orifice at a given point by means of current meters, rod floats, weirs, or other measuring devices.

Gay-Lussac's Law. A law that states that volumes of all gases that react and that are produced during the course of a reaction are related, numerically, to one another as a group of small, whole numbers.

Gel. A two-phase colloidal system consisting of a solid and a liquid in more solid form than a sol.

Gel filtration (molecular exclusion chromatography; molecular sieve chromatography). A type of chromatography which separates molecules on the basis of size, with the higher-molecular-weight substances passing through the column first.

Gel permeation chromatography. Analysis by chromatography in which the stationary phase consists of beads of porous polymeric material; the moving phase is a liquid.

Gel point. The point at which a liquid begins to exhibit elastic properties and increased viscosity.

Generator. Any person whose process produces a hazardous waste in excess of 100 Kg/month or acutely hazardous waste in excess of 1 Kg/month, or whose actions first cause a hazardous waste to become subject to regulation.

Geneva system. International system of nomenclature for organic compounds based on hydrocarbon derivatives in which the names correspond to the longest straight carbon chain in the molecule.

Genotoxic. The ability to interact with and damage the genetic material (DNA) of the cell, often causing a mutation.

Geochemical weathering. The weathering of rocks due to oxidation, reduction, hydration, solution, and hydrolysis.

Geomorphology. The study of land forms.

Geophone. Seismic instrument used to detect vibrations in the ground.

Geophysics. The science of the planet Earth with respect to its structure, composition, and development.

Ghyben-Herzberg principle. Relationship between saline and freshwater that states that the depth to which freshwater extends below sea level is approximately 40 times the height of the water table above sea level.

Gibbs adsorption equation. A formula for a system involving a solvent and a solute, according to which there is an excess surface concentration of solute if the solute decreases the surface tension, and a deficient concentration of solute if the solute increases the surface tension.

Glaciofluvial deposit. A deposit laid down by a stream originating in a glacier.

Glass electrode. An electrode or half cell in which potential measurements are made through a glass membrane that acts as a cation-exchange membrane.

Glass transition (gamma transition; glassy). The change in an amorphous region of a partially crystalline polymer from a viscous or rubbery condition to a hard and relatively brittle one.

Gleization. The reduction of iron under anaerobic soil conditions; the resulting soil often contains a bluish to greenish matrix of colors along with ferric and manganiferous concretions.

Globe valve. A type of stemmed valve that is used for flow control. The valve has a globe-shaped plug that rises or falls vertically when the stem handwheel is rotated.

Glycerol ($CH_2OHCHOHCH_2OH$). A trihedric alcohol that is a completely soluble in water and alcohol but only partially soluble in common solvents such as ether and ethyl acetate. It is used in the manufacture of resins, explosives, antifreezes, medicines, inks, perfumes, cosmetics, soaps, and finishes.

Glycin ($C_8H_9NO_3$). A crystalline compound soluble in alkanes and mineral acids that is used as a photographic developer and in the analytical determination of iron, phosphorous, and silicon.

Glycol. An alcohol that has two hydroxyl groups per molecule.

Gold hydroxide (Au_2O_3). A yellow-brown, light-sensitive, water-insoluble powder that dissolves in most acids and is easily reduced to metallic gold. It is used in medicine, porcelain, goldplating, and daguerreotypes.

Gold tin purple. A brown powder consisting of a mixture of gold chloride and brown tin oxide. It is used in coloring enamels, manufacturing ruby glass, and painting porcelain.

Goodness of fit. Statistical test used to determine the likelihood that sample data have been generated from a population that conforms to a specified type of probability distribution.

Grab sample. A single sample taken at neither a set time nor flow.

Grade. (1) A measure of the quality of an ore; (2) inclination or slope of a stream channel, conduit, or natural ground surface, usually expressed in terms of the ratio or percentage of number of units of vertical rise or fall per unit of horizontal distance.

Gradient. Rate of change of any characteristic per unit of length.

Graduate. A cylindrical vessel that is calibrated in fluid ounces or millimeters to measure the volume of liquids.

Graham's Law. Law that states that the rates of diffusion of gases are inversely proportional to the square roots of their densities.

Grain. Unit of weight equivalent to about 1/15.5 gram or 1/7000 pound which is used in weighing textile fibers and filaments and for specifying the amounts of drugs used in prescriptions and medical preparations.

Grain size. The general dimensions of the particles in a sediment or rock, or of the grains of a particular mineral that make up a sediment or rock. It is common for these dimensions to be referred to with broad terms, such as fine, medium, and coarse. A widely used grain size classification is the Udder-Wentworth grade scale.

Grain size distribution curve. The plot of a sieve analysis for a soil that shows how much of a sample material is smaller or larger than a given particle size. A common plot is with the cumulative percent retained on the y-axis and the grain size plotted on the x-axis.

Gram. Standard unit of weight equivalent to 1/453.59 pound, or 15.4 grains. It is the weight of 1 milliliter of water at 4°C and 1 atmosphere pressure.

Gram atomic weight. Term that refers to the quantity of an element (expressed in grams) that corresponds to the atomic weight.

Gram molecular volume. Volume occupied by a gram molecular weight of a chemical in the gaseous state at 0°C and 760 millimeters of pressure.

Gram molecular weight (GMW). The molecular weight in grams of any particular compound. Synonymous with the term "mole".

Granite (granite gneiss). A coarse-grained rock that contains approximately 25% quartz, 65% or less potassium feldspar, and lesser amounts of mica and hornblende.

Granular media filtration. Physical treatment technique that uses gravity to remove solids from a fluid by passage of the fluid through a bed of granular material.

Granulate. To form or crystallize into grains, granules, or small masses.

Grating spectrograph. A grating spectroscope provided with a photographic camera or other device for recording the spectrum.

Grating spectroscope. A spectroscope which employs a transmission or reflection grating to disperse light. It usually also has a slit and a mirror or lenses to collimate the light sent through the slit to focus the light dispersed by the grating into spectrum lines, as well as an eyepiece for viewing the spectrum.

Graven. A down-dropped area within the slide mass, bounded by two slip surfaces.

Gravimetric. The use of weight as a unit of measurement.

Gravimetric method. A procedure for determining the soil moisture content, which is expressed as a percentage of dry soil weight.

Gravimetric water content. A ratio that expresses the mass of water relative to the mass of dry soil particles; synonymous with the term *mass wetness*.

Gravitational constant (g). A physical constant equal to 9.81 meters/second^2 or 32.2 feet/second^2.

Gravitational convection. Motion caused only by density differences within the fluid.

Gravitational water. Water that moves under force of gravity.

Grease. Group of substances including fats, waxes, free fatty acids, calcium and magnesium soaps, mineral oils, and certain other nonfatty materials.

Grease skimmer. Device for removing floating grease or scum from the surface of wastewater in a tank.

Grease trap. Device for the separation of grease from wastewater by flotation so that it can be removed from the surface.

Green sand. Sand consisting entirely or in large part of particles of the mineral glauconite.

Grid spectrometer. A grating spectrometer in which a large increase in light flux without loss of resolution is achieved by replacing entrance and exit slits with grids consisting of opaque and transparent areas.

Ground-penetrating radar (GPR). A geophysical method used to identify surface formations which will reflect electromagnetic radiation. GPR is useful for defining the boundaries of buried trenches, metal objects, and other subsurface installations on the basis of time-domain reflectrometry.

Groundwater. Subsurface water that occurs beneath the water table and whose pore space is fully saturated.

Groundwater basin. A pervious formation with sides and bottom of relatively impervious material in which groundwater is held or retained.

Groundwater cascade. The descent of groundwater on a steep hydraulic gradient to a lower and flatter water-table slope.

Groundwater divide. Line representing the underground division on a water table or other piezometric surface on either side of which the surface slopes downward.

Groundwater drain. A drain that carries away groundwater.

Groundwater drainage. The outflow or artificial removal of water within a soil, generally accomplished by lowering the water table or preventing its rise.

Groundwater level. The level below which the rock and subsoil are saturated with water.

Groundwater province. An area characterized by a general similarity in the mode of occurrence of groundwater.

Groundwater recharge. Water descending to the zone of saturation from the ground surface.

Groundwater reservoir. A reservoir in which groundwater is stored for future extraction and use.

Groundwater ridge. A ridge-shaped feature of a water table, usually due to excessive seepage of surface water from the area overlying the ridge.

Groundwater storage. Water temporarily stored within the interstices of permeable rocks.

Groundwater table. Surface between the zone of saturation and the zone of aeration; the surface of an unconfined aquifer.

Groundwater velocity. The rate at which groundwater flows through a porous media; it has dimensions of length per unit time. Darcy's Law can be used to approximate groundwater velocity by using the following equation:

$$V = [K(h_1 - h_2)/L]/\eta_e$$

where

K = saturated hydraulic conductivity.
$h_1 - h_2$ = elevational difference between two points.
L = horizontal distance between the two points, h_1 and h_2.
η_e = effective porosity of the porous media.

Grout. A slurry of cement or bentonite that is used to form an impermeable seal within a cavity or in the pore space of a soil. A grout is mixed to a consistency that can be forced through a pipe for placement. Other grout materials include epoxy resins, silicone rubbers, lime, fly ash, and bituminous compounds.

Grout curtain. A barrier to water movement that is formed by injecting a material into the open areas of a rock or soil.

Grout plug. Cement/bentonite mixture used to seal a borehole that has been drilled to a depth greater than the final depth at which the monitoring well is to be installed.

Guard pipe. A pipe, usually made of steel, placed around that portion of the well riser pipe that extends above the ground surface.

H

Half-cell potential. The electrical potential developed by a cell reaction. It can be considered as the sum of the potential developed at the anode and the potential developed at the cathode, each being a half-cell.

Half-life. For radioactive substances, the half-life is the length of time required for half of a given amount of material to disintegrate through radiation. In a chemical sense, the half-life is the time required for one half of a given material to undergo a chemical reaction.

Halocarbon. A compound of carbon and a halogen, sometimes with hydrogen.

Halogen. Any of the elements of the halogen family, consisting of fluorine, chlorine, bromine, iodine, and astatine.

Halogenated fluorocarbons. A generic name for ethane- and methane-based compounds in which some or all of the hydrogen in the compounds are replaced by chorine, bromine, and/or fluorine.

Halogenated hydrocarbon. Organic compound containing one or more halogens (e.g., fluorine, chlorine, bromine, and iodine).

Halogenation. The chemical process or reaction in which a halogen such as fluorine, chlorine, bromine, or iodine is introduced into a substance.

Halon. A halogenated fluorocarbon that contains bromine.

Hardness. The metallic content of water (i.e., positive polyvalent ions, principally calcium and magnesium) which reacts with sodium soaps to produce solid soaps and with negative ions when the water is evaporated in boilers to produce solid boiler scale. Hardness is usually expressed as mg/L of equivalent calcium carbonate ($CaCO_3$).

Hardpan. A dense, hard layer in the subsoil that obstructs penetration of roots and water.

Head. Energy contained in a water mass which is produced by elevation, pressure, or velocity.

Head scarp or main scarp. A steep surface on the undisturbed ground at the top of a slide.

Head space. The air space above a soil or aqueous sample in a closed container into which organic compounds can volatilize. For example, when a volatile organic compound vial is three quarters filled with water or soil, the remaining quarter of the vial is headspace. The air from the headspace (after agitation) is sampled with a portable gas chromatograph. Only gaseous headspace (no liquids) can normally be injected ports.

Head, total (H_T). The *total* head of a liquid at a given point is the sum of three components: (1) *elevation* head, h_o, which is equal to the elevation of the point above a datum; (2) *pressure* head, h_p, which is the height of a column of static water that can be supported by the static pressure at the point; and (3) *velocity* head, h_v, which is the height the kinetic energy of the liquid is capable of lifting the liquid.

Heat capacity. Quantity of energy that must be supplied to raise the temperature of a substance. For contaminated soils, heat capacity is the quantity of energy that must be added to the soil to volatilize organic components. The typical range of heat capacity of soils is relatively narrow; therefore, variations are not likely to have a major impact on application of a thermal desorption process.

Heat exchanger. Device providing for the transfer of heat from a fluid flowing in tubes to another fluid outside the tubes.

Heat of activation. Increase in enthalpy when a substance is transformed from a less active to a more reactive form at constant pressure.

Heat of combustion. Amount of heat released in the oxidation of 1 mole of a substance at a constant pressure or volume.

Heat of crystallization. Heat evolved or absorbed when a crystal forms from a saturated solution of a substance.

Heat of hydration. Heat evolved during the setting and hardening of Portland cement.

Heat of ionization. Increase in enthalpy when 1 mole of a substance is completely ionized at constant pressure.

Heat of reaction. Heat evolved or absorbed when a chemical reaction occurs in which the final state of the system is brought to the same temperature and pressure as that of the initial state of the reacting system.

Heat of solution. Heat evolved or absorbed when a substance is dissolved in a solvent.

Heat of sublimation. Heat required to convert the unit mass of a substance from the solid to the vapor state (sublimation) at a specified temperature and pressure, without the appearance of the liquid state.

Heat transfer. Transfer of heat from one fluid to another.

Heaving sand. Unconsolidated sand that cannot maintain the integrity of the borehole wall and flows into the borehole.

Heavy metals. A group of elements whose compounds are toxic to humans when found in the environment and which have molecular weights. Examples of heavy metals are cadmium, mercury, copper, nickel, chromium, lead, selenium, zinc, arsenic, and plutonium.

Hectare. A metric measure of area equal to 2.47 acres.

Helium spectrometer. A mass spectrometer used to detect the presence of helium in a vacuum system.

Hellige-Truog test. A colorimetric test for soil pH.

Hemiacetal. Class of compounds resulting from the reaction of an aldehyde and alcohol.

Hemic soil (peat or peaty muck). A soil in which 1/3 to 2/3 of the total mass is composed of organic fibers.

Henry's Law. The relationship between the partial pressure of a compound and the equilibrium concentration in the liquid through a proportionality constant known as the Henry's Law constant.

Henry's Law constant (K_H). Sometimes referred to as the air/water partition coefficient, the Henry's Law constant is defined as the ratio of the partial pressure of a compound in air to the concentration of the compound in water at a given temperature under equilibrium conditions. If the vapor pressure and solubility of a compound are known, this parameter can be calculated at 1 atm (760 mmHg) as follows:

$$K_H = PFW/760S$$

where

K_H = Henry's Law constant (atm · m³/mol)
P   = pressure (mm)
FW = formula weight (g/mol)
S   = solubility (mg/L)

Hepatic mixed-function oxidase enzyme activity. Exposure of fish to environmental contaminants such as polynuclear aromatic hydrocarbons (PAHs) and chlorinated hydrocarbons can induce increased activity in enzyme systems capable of detoxifying the contaminants. Hepatic mixed-function oxidase activity is measured as an index of the exposure of fish to contaminants that may harm their reproduction or development.

Hepatopancreas. In zoology, a glandular organ of many invertebrates, usually called the liver.

Heptachlor ($C_{10}H_7C_{17}$). An insecticide that is insoluble in water but soluble in alcohol and xylene.

Heptadecanol ($C_{17}H_{35}OH$). An alcohol that is used as a chemical intermediate, as a perfume fixative, in cosmetics and soaps, and to manufacture surfactants.

Heptaldehyde ($C_6H_{13}CHO$). An aldehyde used as a chemical intermediate and in the manufacture of perfumes and pharmaceuticals.

Heptane ($CH_3(CH_2)_5CH_2$). A hydrocarbon used as an anesthetic and solvent and in standard octane-rating tests.

Heptene ($C_{17}H_{14}$). A liquid that is a mixture of isomers. It is used as an additive in lubricants, as a catalyst, and as a surface-active agent.

Herbicide (defoliant). Chemical used to eliminate plant growth.

Heterocyclic. An unsaturated cyclic compound containing one or more atoms other than carbon as part of the ring structure. The rings may be hexagonal or pentagonal, the latter appearing in the furan, purine, and pyrrole families of heterocyclics.

Heterocyclic PAH. A polynuclear aromatic hydrocarbon compound in which one or more of the ring-bound carbon atoms is replaced by an atom of nitrogen, oxygen, or sulfur.

Heterogeneous. A material whose properties are dependent on its location.

Heterogeneous immunoassay. An immunoassay in which it is not necessary to separate the antigen-antibody complex from the free reactants prior to end-point measurement.

Heterotrophic. Bacteria that oxidize organic matter for energy.

Hexachlorobenzene (C_6Cl_6). A compound consisting of needle-like crystals used in organic synthesis and as a fungicide.

Hexachlorobutadiene ($Cl_2C{:}CClCCl{:}CCl_2$). A colorless liquid soluble in alcohol and ether, insoluble in water, and used as a solvent, heat transfer liquid, and hydraulic fluid.

Hexacontane ($C_{60}H_{122}$). A solid, saturated hydrocarbon of the methane series.

Hexane (C_6H_{14}). A water-insoluble, toxic, flammable, colorless liquid with faint aroma; forms include *n*-hexane, a straight-chain compound that boils at 68.7 and is used as a solvent, paint diluent, alcohol denaturant, and polymerization-reaction medium; isohexane, a mixture of hexane isomers; and neohexane.

Hexavalent chromium (Cr^{VI}). Chromium with a valence of +6.

Hexyl alcohol ($CH_3(CH_2)_4CH_2OH$). A colorless liquid used as a chemical intermediate for pharmaceutical, perfume, esters, and antiseptics manufacture.

High-density polyethylene (HDPE). A polymer produced by the low-pressure polymerization of ethylene as the principal monomer.

Histobar. Similar to a histogram with the exception that the bars are suspended from the best-fitting normal distribution rather than plotted from the horizontal axis.

Histogram. Bar diagram representing a frequency distribution.

Hofmann reaction. Reaction in which amides are degraded by treatment with bromine and alkali (caustic soda) to amines containing one less carbon.

Holocene. A geologic epoch that began 11,000 years ago.

Holocene fault. A fault that has had surface displacement within Holocene time (last 11,000 years).

Homocyclic compound. A ring compound such as benzene that has one type of atom in its structure.

Homogeneity. A material whose properties are identical, regardless of position.

Homogeneous. Uniform in structure or composition regardless of position.

Homologous. In biology, anatomical features of different organisms (species) which correspond in structure and evolutionary origin (e.g., the flipper of a seal and the arms of a human being). In chemistry, the members of a series of organic compounds each having the same structure but differing from the preceding one by a constant increment, such as in the methane series.

Homolysis. The symmetrical breaking of a covalent electron bond.

Hooke's Law. A statement of elastic deformation, in which the strain is linearly proportional to the applied stress.

Horizons, soil. Soils formed in places by internal soil-forming processes.

Hose barb. A twist-type connector used for connecting a small-diameter hose to a valve or faucet.

Hue. Color.

Humus. Biologically stable organic matter that is the product of aerobic decomposition of the vegetable tissues of plants.

Hydration. The chemical process of combination or union of water with other substances.

Hydraulic conductivity, (K). The proportionality factor in Darcy's Law that applies to the viscous flow of water in soil (i.e., the flux of water per unit gradient of hydraulic potential). Hydraulic gradient is measured as the rate of flow of water in gallons per day through a cross-section of one square foot under a unit hydraulic gradient at the prevailing temperature (gpd/ft^2). In the SI system, the units are $m^3/day/m^2$ or m/day.

Hydraulic continuity. A water bridge or connection between two or more geological formations.

Hydraulic diffusivity (D). The conductivity of the saturated medium divided by the specific storage.

Hydraulic fill. An earth structure or grading operation in which the fill material is transported and deposited by means of water pumped through a flexible or rigid pipe.

Hydraulic fracturing. A technique commonly used to increase the yields of oil wells. The technique involved injecting a fluid into a well until the pressure of the fluid exceeds a critical value and a fracture is nucleated. A granular material is then pumped into the fracture.

Hydraulic gradient. The change in static head per unit of distance in a given direction. If not specified, the direction is understood to be that of the maximum rate of decrease in head. The gradient of the head is a mathematical term which refers to the vector denoted by Δh, the magnitude of which (dh/dl) is equal to the maximum rate of change in head and the direction of which is that in which the maximum rate of increase occurs. The hydraulic gradient and the gradient of the head are equal but of opposite sign.

Hydraulic head. The height of a free surface of a water body above a given point beneath the surface.

Hydraulic loss. The loss of head attributable to obstructions, friction, or changes in velocity.

Hydraulics. That branch of science or engineering which deals with water or other fluid in motion.

Hydrazine (N_2NNH_2). A liquid used as a rocket fuel, in corrosion inhibition in boilers, in the synthesis of biologically active materials, explosives, antioxidants, and photographic chemicals.

Hydrazine hydrate ($H_2NNH_2OH_2O$). A fuming liquid used as a component in jet fuels and as an intermediate in organic synthesis.

Hydride. A compound, such as hydrogen sulfide, containing hydrogen and another element.

Hydrocarbons. A large and important group of organic compounds that contain only hydrogen and carbon. There are two types: saturated and unsaturated. Saturated hydrocarbons are those in which adjacent carbon atoms are joined by a single valence bond and all other valences are satisfied by hydrogen. The saturated hydrocarbons form a whole series of compounds starting with one carbon atom and increasing by one carbon atom, stepwise. These compounds are also known as the paraffin series, the methane series, and the alkanes. The principal source is petroleum. Gasoline is a mixture containing several of them; diesel fuel is another such mixture. The unsaturated hydrocarbons are usually separated into four classes: (1) the ethylene series of compounds, all containing one double valence bond between two adjacent carbon atoms; (2) the diolefin series of compounds, all containing two double bonds in their molecules; (3) the polyenes, containing more than two double bonds; and (4) the acetylene series of unsaturated hydrocarbons having a triple bond between adjacent carbon atoms. These compounds are found in some industrial wastewater, particularly those from the manufacture of some types of synthetic rubbers.

Hydrochloric acid (HCl). A solution of hydrogen chloride gas in water; a poisonous, pungent liquid forming a concentration in water which is used as a reagent and in organic synthesis, acidizing oil wells, ore reduction, food processing, metal cleaning, and pickling.

Hydrocompaction. The process whereby soils collapse when wetted.

Hydrocyanic acid (HCN). A highly toxic liquid that has the odor of bitter almonds and is used to manufacture cyanide salts, acrylonitrile, and dyes and as a fumigant in agriculture.

Hydrodynamic dispersion. The spreading of groundwater during transport that is due primarily to the processes of mechanical mixing and molecular diffusion.

Hydrofluoric acid (HF). An aqueous solution of hydrogen fluoride. It is colorless, fuming, poisonous, and extremely corrosive. It will attack glass and other silica materials and is used to polish, frost, and etch glass; to pickle copper, brass, and alloy steels; to clean stone and brick; to etch semiconductor wafers; to acidize oil wells; and to dissolve ores.

Hydrogel. The formation of a colloid in which the disperse phase (colloid) has combined with the continuous phase (water) to produce a viscous jelly-like product.

Hydrogen bonding. The bonding of two electronegative atoms by a hydrogen atom. The hydrogen atom is linked to one electronegative atom by a covalent bond and to the other by an electrostatic bond.

Hydrogen bromide (HBr). A hazardous, toxic gas used as a chemical intermediate and as an alkylation catalyst.

Hydrogen ion concentration. The normality of a solution with respect to hydrogen ions.

Hydrogen line. A spectral line emitted by neutral hydrogen having a frequency of 1420 megahertz and a wavelength of 21 centimeters.

Hydrogen peroxide (H_2O_2). Also known as peroxide. An unstable, colorless liquid that is soluble in water and alcohols. Hydrogen peroxide is used extensively in bleaching cotton, wool, silk, rayon, linen, paper pulp, and other fibrous materials. As an oxidizing agent, it is employed in the manufacture of niacin, dyes, rocket fuel, drugs, and pharmaceuticals. Dilute solutions of hydrogen peroxide have long been used in the treatment of open wounds.

Hydrogen selenide (H_2Se). A flammable, toxic gas that is soluble in water, carbon disulfide, and phosgene. It is used to make metallic selenides and organoselenium compounds and in the preparation of semiconductor materials.

Hydrogen sulfide (H_2S). A flammable, toxic gas that is soluble in water and alcohol and is used as an analytical reagent, as a sulfur source, and for purification of hydrochloric acid.

Hydrogenation. Chemical combination of hydrogen with another substance via heat and pressure in the presence of a catalyst.

Hydrogenesis. Natural condensation of moisture in the air spaces in the surface soil or rock.

Hydrogenolysis. A reduction process in which a carbon halogen bond is broken and hydrogen replaces the halogen substituent.

Hydrogeology. The branch of hydrology that deals with groundwater.

Hydrograph. The graphical plot of changes in water flow or water level vs. time.

Hydrologic budget equation (water balance). An equation that assumes that the input minus the output of a hydrologic system is equal to the change in storage of that system. The change in storage is usually reported in units of acre feet per year.

Hydrologic cycle. The cycle that is described by the endless circulation of water in the atmosphere, on the ground surface, below the surface, and in the atmosphere.

Hydrology. The science dealing with the properties, distribution, and circulation of water.

Hydrolysis. (1) Chemical reaction in which water reacts with another compound so that both it and the compound are split; the water decomposes to its ions, hydrogen and hydroxyl, each of which reacts with a portion of the split compound. (2) Chemical reaction in which an organic chemical reacts with water or a hydroxide ion.

Hydrolytic process. A reaction of both organic and inorganic chemistry wherein water effects a double decomposition with another compound, hydrogen going to one compound and hydroxyl to another.

Hydrolyze. To subject to hydrolysis; to undergo hydrolysis.

Hydromechanics. The science that studies the equilibrium and motion of fluids and of bodies in or surrounding them.

Hydrometer. An instrument designed to measure the specific gravity or weight per unit volume of a fluid by the depth to which it sinks in the fluid.

Hydrophile-lipophile balance. The relative simultaneous attraction of an emulsifier for two phases of an emulsion system (for example, water and oil).

Hydrophilic. A substance that absorbs or adsorbs water.

Hydrophobic. A substance that repels water; tending not to combine with water or incapable of dissolving in water; insoluble or immiscible in water. A property exhibited by nonpolar organic compounds, including the petroleum hydrocarbons.

Hydrophobicity (lipophilicity). In chemistry, the preferential migration and accumulation of an organic chemical in hydrophobic solvents or on a hydrophobic surface.

Hydropunch®. A groundwater sampling tool that can be used with a cone penetrometer or conventional drilling equipment. The Hydropunch® consists of a stainless steel drive point, a perforated section of stainless steel pipe for sample intake, a stainless and Teflon® sample chamber that is capable of collecting 500 mL of groundwater, and an adapter to attach the unit to either penetrometer push rods or standard soil sampling drill rods.

Hydroquinone ($C_6H_4(OH)_2$). White crystals which are soluble in alcohol, ether, and water. It is used in photographic dye chemicals, in medicine, as an oxidant and inhibitor, and in paints, varnishes, motor fuels, and oils.

Hydroquinone monomethyl ether ($CH_3OC_6H_4OH$). A white, waxy solid that is soluble in benzene, acetone, and alcohol. It is used in the manufacture of antioxidants, pharmaceuticals, and dyestuffs.

Hydrosol. A colloidal system in which the dispersion medium is water and the dispersed phase is a solid, gas, or another liquid.

Hydrothermal. Pertaining to or resulting from the activity of hot aqueous solutions originating from magma or other source deep in the earth.

Hydrotrope. A compound with the ability to increase the solubilities of slightly soluble organic compounds.

Hydrous. Indicating the presence of an indefinite amount of water.

Hydroxide. A compound containing the OH⁻ group.

Hydroxide alkalinity. Alkalinity caused by hydroxyl ions.

Hydroxy acid. Acids that have OH groups attached to the molecule other than in the carboxyl group. Chemically, they act as acids and alcohols.

Hydroxylation. The addition of a hydroxyl group.

Hygrometer. An instrument for measuring the relative amount of moisture in the atmosphere.

Hygrometric (dew point) method. Technique used to measure the water potential by determining the dew point depression temperature.

Hygroscopic. Water that possesses the ability to accelerate the condensation of water vapor. Pertaining to water absorbed by dry soil minerals from the atmosphere; the amounts depend on the physicochemical character of the surfaces and increase with rising relative humidity.

Hygroscopic water. Water molecules held on surfaces of particles by forces of adhesion.

Hygroscopicity. As applied to soil, the ability to absorb and retain moisture.

Hyperfine structure. A splitting of spectral lines due to the spin of the atomic nucleus or to the occurrence of a mixture of isotopes in the element.

Hyperthermic soil. A soil with a temperature greater than 22°C.

Hypochlorite (ClO_3). A negative ion derived from hypochlorous acid which is used as an oxidizing agent and a constituent of bleaching agents.

Hypoiodous acid (HIO). A very weak unstable acid that occurs as the result of the weak hydrolysis of iodine in water.

Hypothermal ore deposits. Ore deposit formed at high temperatures (300 to 500°C) and pressures. Most ore deposits formed under these conditions are coarse grained. Examples are fluorite, wolframite, barite, galena, stannite, pyrite, and uraninite.

Hypothesis. In statistics, a formal statement about a parameter of interest and the distribution of a statistic.

Hypoxic. A condition of low oxygen concentration, below that considered to be aerobic.

Hysteresis. A change in the shape of a soil water characteristic curve which differs depending upon whether soil sorption (wetting) or desorption (draining) occurs.

I

Ice point. The true freezing point of water at which a mixture of air-saturated pure water and pure ice may exist in equilibrium at a pressure of 1 standard atmosphere.

I.D./O.D. Inside diameter and outside diameter of a container, respectively.

Ideal gas constant. A proportionality constant with a numerical value depending on the units in which pressure and volume are measured. If pressure is expressed in atmospheres and volumes in liters, then $R = 0.082054$ liter atm $deg^{-1}mol^{-1}$.

Igneous rocks. Rocks that solidified from molten or partly molten material (i.e., magma).

Ignitabiltiy. A characteristic of a hazardous waste whereby it is easily combustible or flammable.

Ignition. The process of starting a fuel mixture burning or the means for such a process.

Illite. General name for a group of three-layer, mica-like clay minerals. These clay minerals are intermediate in composition and structure (between muscovite and montmorillonite). A moderate to high concentration of silica and aluminum is required for stability.

Illuvial iron. Iron that appears as coatings on sand or silt particles.

Illuviation. The movement of material into a portion of the soil profile from an overlying horizon.

Imhoff cone. A 1-liter container that is used to estimate the concentration of suspended sediment in a water sample.

Imhoff tank. A deep, two-storied wastewater tank that consists of an upper, continuous-flow sedimentation chamber and a lower, sludge-digestion chamber.

Imine. A class of compounds that are the product of condensation reactions of aldehydes or ketones with ammonia or amines.

Immersion sampling. The collection of a liquid sample for laboratory or other analysis by immersing a container in the liquid and filling it.

Immiscible. Liquids that do not mix.

Immunoassay. An assay procedure based on the reversible and noncovalent binding of an antigen by an antibody using a labeled form of one or the other to quantify the system. It can be used to detect or quantify either antigens or antibodies.

Impairment. A change in water quality which makes it less suitable for beneficial use.

Impeller pump. Any pump wherein the water is moved by the continuous application of power derived from some mechanical agency or medium.

In-line rotameter. A flow-measurement device for liquids and gases that uses a flow tube and specialized float. The float device is supported by the flowing fluid in the clear glass or plastic flow tube. The vertical-scaled flow tube is calibrated for the desired flow volumes per unit of time.

In situ. In its original place; unmoved; unexcavated; remaining in the subsurface.

In situ **biodegradation.** An *in situ* treatment technology that enhances the ability of a microorganism to degrade organic compounds in the soil or groundwater.

In situ **soil treatment**. Treatment of contaminated soils without excavation.

Incinerator. Any enclosed device using controlled flame combustion that neither meets the criteria for classification as a boiler nor is listed as an industrial furnace.

Inclination. A general term for the measured vertical angle between the horizontal and a plane or line.

Inclinometer. An instrument used to determine if any movements have taken place within a slope. The instrument consists of a pendulum that indicates deviations from the vertical of the shape and the position of flexible vertical casings.

Incubation. The maintenance of chemical mixtures at specified temperatures for varying time periods for the purpose of studying chemical reactions, such as enzyme activity.

Indan ($C_6H_4(CH_2)_3$). A colorless liquid that is soluble in alcohol and ether and insoluble in water. It is derived from coal tar.

Indanthrene ($C_{28}H_{14}N_2O_4$). A blue pigment or vat dye that is used in cotton dyeing and as a pigment in paints and enamels.

Indene (C_9H_8). A liquid, polynuclear hydrocarbon derived from coal tar distillates.

Indicator parameters. Parameters that are used as an indicator for the presence of other compounds. Examples include pH, specific conductance, total organic carbon (TOC), and total organic halogens (TOX).

Indigenous. Living or occurring naturally in a specific area or environment; native.

Indium arsenide (InAs). Metallic crystals that are used as an inter-metallic compounds with semiconductor properties.

Induction period. Time of acceleration of a chemical reaction from zero to a maximum rate.

Indurated soil. Soil cemented into a hard mass that will not soften when wetted.

Induration. The geologic process of hardening of sediments or other rock aggregates through cementation, pressure, heat, or other causes.

Inert ingredient. A component of a pesticide such as a solvent or carrier that is not active against target pests.

Infectious. Capacity of a microorganism to invade a susceptible host, replicating and causing an altered host reaction.

Infiltration. Flow or movement of water through the interstices or pores of a soil or other porous medium.

Infiltration coefficient. Ratio of infiltration to precipitation.

Infiltration gallery. A sizable gallery with openings in its sides and bottom which extends generally horizontally into a water-bearing formation.

Infiltration rate. Rate at which water flows into soil or other porous material per unit surface area.

Infrared film. Film that is primarily sensitive to the blue-violet, red, and near infrared radiations.

Infrared radiation. Radiation whose wavelengths are in the infrared radiation spectrum; wavelength greater than 7600 angstroms.

Infrared spectrometer. A device used to identify and measure the concentration of heteroatomic compounds in gases, nonaqueous liquids, and some solids.

Infrared spectrophotometry. Measurement of light absorption in the infrared region.

Inhibitor. A substance that retards or reduces the rate of a chemical reaction.

Initiator. A substance or molecule that initiates a chain reaction; polymerization is an example.

Injection well. A well used to inject under pressure a fluid (liquid or gas) into the subsurface.

Inlet well. A well through which a fluid (liquid or gas) is allowed to enter the subsurface under natural pressure.

Inoculate. To implant microorganisms onto or into a culture medium.

Inorganic. Matter that is mineral in origin and does not contain carbon.

Inorganic acid. A compound composed of hydrogen and a nonmetal element or radical such as hydrochloric acid (HCl), sulfuric acid (H_2SO_4), or carbonic acid (H_2CO_3).

Inorganic matter. Chemical substances of mineral origin, not containing carbon-to-carbon bonding.

Inorganic pigment. A natural or synthetic metal oxide, sulfide, or other salt that is used as a coloring agent for paints, plastics, and inks.

Interceptor trench. A groundwater remediation technology whereby a trench or ditch is constructed across the leading edge of a contaminant plume. Contaminants are then pumped or skimmed from the groundwater as it enters the trench.

Interdiffusion. The self-mixing of two fluids, initially separated by a diaphragm.

Interface. The boundary between any two phases such as gas/liquid, gas/solid, liquid/liquid, liquid/solid, and solid/solid.

Interface mixing. The mixing of two immiscible or partially miscible liquids at the plane of contact (interface).

Interference. A situation that arises when a foreign substructure is affected in any way by a direct current source.

Interference spectrum. A spectrum that results from the interference of light, as in a very thin film.

Interfingering. In soil science, a diagnostic term used to describe a soil that contains narrow (less than 5 millimeters) intrusions of a light-colored material into a clay or a clay layer whose cation exchange capacity consists of at least 15% sodium base saturated.

Intergranular. Between the individual grains in a rock or sediment.

Intermediate. A precursor to a desired product.

Interpolymer. A mixed polymer made from two or more starting materials.

Interstice. A pore or open space in rock or granular material, not occupied by solid matter.

Interstitial water (pore water). Water that occupies an open space between solid soil particles.

Intrinsic permeability (k). A variable that describes the Darcy's "K" or saturated hydraulic conductivity. It is a function of the medium and has dimensions of length squared. It is equal to:

$$k = K\mu/\rho g$$

where

> K = saturated hydraulic conductivity.
> μ = dynamic viscosity.
> ρ = fluid density.
> g = acceleration due to gravity

Intrinsic tracer. An isotope that is naturally present in a form suitable for tracing a given element through chemical and physical processes.

Intrusive rocks. Rocks formed from magma injected beneath the Earth's surface. These rocks generally have large crystals in their matrix.

Invar®. An alloy with a very low coefficient of expansion at atmospheric temperatures; it consists of 36% nickel and 0.35% manganese, with the remainder being an iron and carbon mixture.

Inverse distance method. Method used to grid data in which the data points are weighted such that the influence of one data point on another declines with distance from the point being estimated. The greater the weighing power, the faster the decline in influence and the less effect points farther out will have on the interpolation.

Ion exchange. Removal of ions from a solution through the use of an ion exchange medium.

Ion-exchange capacity. The measured ability of a formation or soil to adsorb charged atoms or molecules.

Ion-exchange chromatography. A chromatographic procedure in which the stationary phase consists of ion-exchange resins.

Ion-exchange treatment. The use of ion-exchange materials such as resin or zeolites to remove undesirable ions from a liquid and substitute acceptable ions.

Ion exclusion. An ion-exchange resin system in which mobile ions in the resin-gel phase electrically neutralize the immobilized charged functional groups attached to the resin.

Ion-selective electrode. An electrode that has a high degree of selectivity for one ion in a solution.

Ion kinetic energy spectrometry. A spectrometric technique that uses a beam of ions of high kinetic energy passing through a field-free reaction chamber from which ionic products are collected and energy analyzed.

Ion scattering spectroscopy. A spectroscopic technique in which a low-energy beam of inert-gas ions is directed at a surface, and the energies and scattering angles of the scattered ions are used to identify surface atoms.

Ionic bond (electrovalent bond). A type of chemical bond in which atoms of different elements unite by transferring one or more electrons from one atom to the other to form an ionizing or polar compound.

Ionic concentration. The concentration of any ion in solution, generally expressed in moles per liter.

Ionic equilibrium. The condition in which the rate of dissociation of non-ionized molecules is equal to the rate of combination of the ions.

Ionic strength (μ). An empirical measurement of the interactions among ions in a solution. The ionic strength of a solution is defined as:

$$\mu = {}^{1/2}\Sigma C_i Z_i^2$$

where

μ = ionic strength of the solution.
${}^{1/2}\Sigma$ = summation of the ionic strengths and concentrations in the solution.
C_i = molar concentration of the ion.
Z_i^2 = charge of the ion.

Ionization. The process by which a neutral atom or molecule loses or gains electrons, thereby acquiring a net charge and becoming an ion.

Ionization potential (IP). The energy required to remove a given electron from the molecule's atomic orbit (outermost shell) to an infinite distance. It is expressed in electron volts (eV); one electron volt is equivalent to 23,053 cal/mol.

Iron bacteria. Bacteria that assimilate iron and excrete its compounds.

Iron oxide. Any of a number of iron-oxygen compounds in which the iron appears in several oxidation states.

Iron vitriol. A ferrous sulfate.

Isobar. A line of equal barometric pressure, usually referenced to sea level.

Isobutyl alcohol ($(CH_3)_2CHCH_2OH$). A colorless liquid that is a byproduct of the synthetic production of methanol.

Isocetyl laurate ($C_{11}H_{23}COOC_{16}H_{33}$). An oily, combustible liquid that is soluble in most organic solvents. It is used in cosmetics and pharmaceuticals and as a plasticizer and textile softener.

Isoelectric point. (1) Point of electrical neutrality; (2) pH at which the positive and negative charges on the particles of a colloidal solution cancel each other so that the particles become electrically neutral.

Isoelectronic. Pertaining to atoms having the same number of electrons outside the nucleus of the atom.

Isohyet. A line or contour representing equal amounts of precipitation during a given time period or for a particular storm.

Isohyetal map. A map on which precipitation is plotted by connecting points of equal precipitation (isohyetal lines) and which shows rainfall distribution in the area mapped.

Isolation. The separation of a pure chemical substance from a compound or mixture, such as in the distillation, precipitation, or absorption reaction.

Isolation/containment. The use of concrete or asphalt surface covers to isolate or contain chemical compounds in soil.

Isomerism. Phenomenon whereby certain chemical compounds have structures that are different although the compounds possess the same elemental composition.

Isomers. Chemical substances having the same molecular weight and percentage of elements but differing in structure and, therefore, physical and chemical properties.

Isonival. A line of equal snow depth or equal water content of snow.

Isooctane ($(CH_3)_2CHCH_2(CH_3)_3$). A flammable liquid used in motor fuels and as a chemical intermediate.

Isopentanoic acid (C_4H_9COOH). A combustible liquid used for the manufacture of plasticizers, pharmaceuticals, and synthetic lubricants.

Isopleth. A line of equal or constant value of a given quantity, with respect to either space or time.

Isopycnic. A line of equal or constant density, with respect to either space or time.

Isostasy. The balance achieved between plates in the Earth's crust.

Isosteric. Referring to similar electronic arrangements in chemical compounds.

Isotatic. (1) Subject to equal pressure from every side; (2) crystalline polymers in which substituents in the asymmetric carbon atoms have the same (rather than random) configuration in relation to the main chain.

Isotherm. A line drawn through all points having the same temperature.

Isotone. One of several nuclides having the same atomic number but different mass number.

Isotope. One of two or more atoms having the same atomic number but different mass number.

Isotope effect. The effect of differences in mass between isotopes of the same element on non-nuclear physical and chemical properties.

Isotope shift. A displacement in spectral lines due to different isotopes of an element.

Isotropic soil. A soil that exhibits identical properties in all directions from a given point.

Isotropy. Condition in which all significant properties are independent of direction. Although no aquifers are truly isotropic in nature, models based upon the assumption of isotropy are valuable for predicting various relationships.

IUPAC. Abbreviation for the International Union of Pure and Applied Chemistry, which is an organization devoted to establishing worldwide uniformity in such matters as chemical units of weight and measurement, symbols, and terminology.

J

Jacob straight-line method. A mathematical method used to analyze aquifer test data to determine aquifer transmissivity and storativity. Drawdown data are plotted on semi-log paper as a function of time since pumping commenced. The equations for storativity and transmissivity are

$$S = T \, t_o/640 \, r^2$$

$$T = 35Q/ds$$

where

S   = aquifer storativity.
T   = transmissivity.
t_o = intercept of the line with the time axis on the graph
r^2 = distance from the pumping well to the observation well.
Q   = pumping rate.
ds = change in the drawdown per one log cycle.

Jetting. A method of well drilling wherein casing is sunk by driving while the material inside is washed out by a water jet and carried to the top of the casing.

Joint. A surface of a fracture or a parting in a rock, without displacement, that divides a rock and along which there has been no visible movement parallel to the plane or surface.

Joint plane. In soil micromorphology, a plane that is more or less parallel in orientation.

Joule (J). A measurement of electrical energy equal to the energy generated by the flow of 1 ampere in 1 second against an electromagnetic force of 1 volt. It is equal to 1 watt second or 0.239 calories.

JP-5 (Navy jet fuel). A mixture of hydrocarbons similar in composition to refined kerosene products. The predominant constituents are compounds containing from 9 to 16 carbon atoms, olefins, naphthionic, and aromatic hydrocarbons. The bulk of the constituents in JP-5 are alkanes and cycloalkanes (75 to 90%).

K

Kaolin. A special type of clay, high in aluminum content.

Kaolinite. A clay mineral with equal concentrations of silica and aluminum.

Karbutilate ($C_{14}H_{21}N_3O_3$). A white solid material used as an herbicide. Also known as meta-(3,3-dimethylureido) phenyl-tert-butylcarbamate.

Karst topography (Karst). Topographic area created by the dissolution of a carbonate rock terrain. This type of topography is characterized by sinkholes, caverns, and lack of surface streams.

Kelvin (K). A unit of temperature equal to 1°C and 1.8°F.

Kelvin temperature scale. A temperature scale in which the freezing point of water is 273.15 K and the boiling point of water is 373.15 K.

Ketones. An alcohol that is prepared by the oxidation of secondary alcohols; acetone is an example of a ketone.

Kettle hole. A depression found in glacial drift created when a block of ice, isolated by the general melting of the glacier, is partly buried by sediments and later melts.

Kinematic viscosity (v). A fluid property that is equal to the viscosity divided by the fluid density, or:

$$v = \mu/\rho$$

where

v = kinematic viscosity.
μ = viscosity.
ρ = fluid density.

Kinematic viscosity coefficient. Ratio of the coefficient of absolute viscosity of a fluid to its unit weight.

Kinetic energy. The energy possessed by a body of matter as a result of its motion.

Kinetic-friction coefficient. A numerical quantity used as an index of the amount of force necessary to keep a body sliding at a uniform velocity on the surface of another body.

Kinetic theory. A theory that states that liquids as well as gases are in constant agitation.

Konimeter. Instrument used to determine the dust content of air sample.

Kriging. A statistical interpolation method that selects the best linear estimate for the variable in question. The variable is assumed to be a random function whose spatial correlation is defined by a variogram.

Kroll process. The process for the production of pure titanium metal by the reduction of titanium tetrachloride with molten magnesium.

L

Laboratory control standard. A sample composed of laboratory reagent water spiked with known compounds and subjected to the same analytical procedures as the sample(s). This procedure indicates the accuracy of the analytical method and, because it is prepared from a different source than the standard used to calibrate the instrument, it serves to verify the equipment calibration.

Lambert's Law. Law dealing with the absorption of light which states that the depth or thickness of a colored liquid absorbs an equal fraction of light that passes through it.

Lamellae. The stacked composite layers of a clay particle.

Laminar flow. Flow of a viscous fluid in which particles of the fluid move in parallel layers, each of which has a constant velocity but is in motion relative to its neighboring layers. Also called straight-line, streamline, or viscous flow.

Land treatment. A waste management system in which wastes are deposited and worked into the soil to allow soil microorganisms to degrade and demobilize the waste within the soil; also known as land spreading, refuse farming, and sludge farming.

Landelier's index. Hydrogen-ion concentration that a water requires to be in equilibrium with its calcium carbonate content.

Landfill. A disposal facility where waste is placed in containers or in bulk form and covered with soil and left in place.

Landform. The physical expression of the land surface.

Landslide. The movement of a mass of earth or rock, or a mixture of both, in a downward direction by sliding.

Langley. A unit of energy equal to 1 gram calorie per square centimeter.

Laser spectroscopy. A branch of spectroscopy in which a laser is used as an intense, monochromatic light source.

Latent energy. Energy required to cause a change of state at constant temperature, as in the melting of ice or the vaporization of water.

Latent heat. The quantity of energy, in calories per gram, absorbed or given off as a substance undergoes a change of state. Examples are when a material changes from a liquid to solid (freezes), from a solid to liquid (melts), from a liquid to vapor (boils), or from a vapor to liquid (condenses).

Lateral spread. The movement of a fractured mass laterally, often along a basal shear surface or zone of plastic flow.

Lateritic. An extreme type of weathering common in tropical climates in which iron and aluminum silicates are decomposed and silica (along with most other elements) are removed by leaching. The product, laterite, is characterized by a high content of alumina and/or ferric oxide.

Laterization. A weathering process involving the increase of alumina or iron oxides, or both, and the removal of silica from a soil.

Latex. An aqueous suspension of proteins and resins occurring in some plants, trees, and shrubs from which natural rubber latex is obtained.

Lauryl mercaptan ($C_{12}H_{25}SH$). A pale-yellow or water-white liquid which is insoluble in water and soluble in organic solvents. It is used to manufacture plastics, pharmaceuticals, insecticides, fungicides, and elastomers.

Law of corresponding states. Law stating that, for two substances, if any two ratios of pressure, temperature, or volume to their respective critical properties are equal, the third ratio must equal the other two.

Law of mass action. Law stating that the rate at which a chemical reaction proceeds is directly proportional to the molecular concentrations of the reacting compounds.

Laws of thermodynamics. Laws that are assumed to apply to all known phenomena. The first law of thermodynamics states that energy can be neither created nor destroyed, and the second law states that heat flows from a hotter to a colder object, but not visa versa.

LC_{50}. The concentration of toxicant in a given vehicle (usually air or water) that is lethal to 50% of the organisms tested in a specified time under specified test conditions.

LD_{50}. The dose of a toxicant that is lethal to 50% of the organisms tested in a specified time under specified test conditions; usually expressed as weight of toxicant per unit body weight of the test organism.

Leachate. A solution produced by the movement or percolation of liquid through soil or solid waste and the subsequent dissolution of certain constituents in the water.

Leaching. The removal of soluble constituents from soils or other material by percolating fluid.

Leaching field. A network of porous pipes and surrounding soil from which septic tank effluent is slowly released. The effluent is gradually decomposed and recycled by natural processes before the effluent reaches a body of ground or surface water.

Lead (Pb). Soft, malleable, ductile, bluish-white, dense metallic element which exists as a variety of toxic salts.

Lead chromate ($PbCrO_4$). Poisonous brown crystals which decompose when heated, are insoluble in water and alcohol, and soluble in glacial acetic acid. It is used as an oxidizing agent; in electrodes, batteries, matches, and explosives; as a textile mordant; in dye manufacture; and as an analytical reagent. Also known as anhydrous plumbic acid, brown lead oxide, and lead peroxide.

Lead fluoride (PbF$_2$). A crystalline solid material used for laser crystals and electronic and optical applications.

Lead iodide (PbI$_2$). Poisonous, water- and alcohol-insoluble golden yellow crystals used in photography, medicine, printing, mosaic gold, and bronzing.

Lead phosphate (Pb$_3$PO$_4$). A poisonous white powder which is soluble in nitric acid and is used as a stabilizer in plastics.

Lead stearate (Pb(C$_{18}$H$_{35}$O$_2$)$_2$). A poisonous white powder which is soluble in alcohol and ether and insoluble in water. It is used as a lacquer and a varnish drier and in high-pressure lubricants.

Lead titanate (PbTiO$_3$). A water-insoluble, pale yellow solid.

Leakage. Phenomenon occurring in an ion-exchange process in which some influent ions are not adsorbed by the ion-exchange bed and appear in the effluent.

Leaky aquifer. An aquifer in which the draw-down in the well produces flow into or out of the aquifer through leaky (semi-confining) layers.

Lessivage. The washing in suspension of fine clays and silt down cracks in a soil.

Lewis base. A substance, such as the hydroxide and ammonia ion, that can donate an electron pair.

Lifetime average daily dose (LADD). A measurement of dose that is usually used in carcinogen risk assessment; it is equal to the maximum daily dose that an individual is likely to receive on any day during the period of exposure multiplied by the fraction of the total lifetime that the individual is exposed to the substance.

Ligand. (1) A soluble molecule or ion that can form complexes with a metal; (2) molecule, ion, or atom that is capable of furnishing or donating one or more pairs of electrons to a transition-metal ion, thus forming a coordination compound.

Ligand membrane. A solvent immiscible with water and a reagent and acting as an extractant and complexing agent for an ion.

Lignin. An organic substance that is present in the cell walls and cellulose fibers of plants.

Limestone. A sedimentary rock with a composition consisting of more than 50% calcium carbonate.

Limnic soil. Organic or inorganic materials deposited in water by the activity of aquatic organisms or derived from underwater and floating organisms; includes diatomaceous earth, marl, and sedimentary peat.

Limnology. The study of freshwater bodies.

Lindane. An insecticide with many applications, including its use as a seed and soil treatment; foliage application on fruit and nut trees, vegetables, and ornamentals; and timber and wood protection. It possesses more vapor activity than most of the organochlorine insecticides.

Lineaments. Straight or gently curved, lengthy features of the Earth's surface, frequently expressed topographically as depressions or lines of depressions.

Linear molecule. A molecule in which the atoms are arranged so that the bond angle between each is 180°.

Linear polymer. A polymer in which molecule is arranged in a chain-like fashion.

Lineation. General, non-genetic term for any linear rock structure. Lineation in metamorphic rocks includes mineral streaking and stretching, crinkles and minute folds parallel to fold axes, and lines of intersection between bedding and cleavage or of variously oriented cleavages.

Liner. A continuous layer of natural or manmade materials lining the bottom and/or sides of a surface impoundment, landfill, or landfill cell that restricts the downward or lateral escape of liquid.

Lipid. Any of a diverse group of organic compounds, occurring in living organisms, that are insoluble in water but soluble in organic solvents, such as chloroform and benzene. Lipids are broadly classified into two categories: complex lipids, which are esters of long-chain fatty acids and include the glycerides (which constitute the fats and oils of animals and plants), glycolipids, phospholipids, and waxes; and simple lipids, which do not contain fatty acids and include the steroids and terpenes.

Lipophilic. A substance with a strong affinity for fats.

Liquefaction. Sudden decrease of shearing resistance of a cohesionless soil, caused by a collapse of the structure by shock or strain and associated with a sudden but temporary increase of the pore water pressure. It involves the transformation of the material into a fluid mass.

Liquefied petroleum gas. A group of hydrocarbon fuels which constitute a portion of the broader group of compressed gases. Liquefied petroleum includes butane, butenes, isobutane, propane, and propylene.

Liquid. A substance that flows freely; a fluid, which is a state of matter intermediate between gaseous and solid. Liquids differ from gases because of their greater density and more complex molecules and from solids because of their random (amorphous) molecular arrangement.

Liquid chromatography. A form of chromatography employing a liquid as the moving phase and a solid or liquid on a solid support as the stationary phase. Examples include column chromatography, gel permeation chromatography, and partition chromatography.

Liquid hydrocarbon. A hydrocarbon that has been converted from a gas to a liquid by pressure or by reduction in temperature.

Liquid index (LI). Quantitative value used to assess whether a soil will behave as a brittle solid, semisolid, plastic, or liquid. LI is equal to the difference between the natural moisture content of the soil and the plastic limit (PL) divided by the plasticity index (PI).

Liquid injection incineration. The process that uses a series of atomizing devices to introduce finely divided droplets of waste mixed with air into a refractory-lined combustion chamber.

Liquid limit (LL). In soil mechanics, the boundary between the liquid and plastic state of a soil.

Liquinox. A laboratory-grade soap mixture often used to clean field equipment used in hazardous waste investigations.

Liquor. In chemical technology, any aqueous solution of one or more chemical compounds.

Liter. A standard unit of volume for gases and liquids; it is the volume occupied by 1000 grams of water at a pressure of 1 atmosphere at 4°C (39°F) and is equivalent to about 1.05 quarts.

Lithium stearate ($LiC_{18}H_{35}O_2$). A white, crystalline compound used in cosmetics, plastics, greases, and as a corrosion inhibitor in petroleum.

Lithium titanate (Li_2TiO_3). A water-insoluble white powder used in titanium-containing enamels and as a mill additive in vitreous and semi-vitreous glazes.

Lithography. Transfer of a pattern or image from one medium to another, as from mask to a wafer.

Lithology. The description of rocks in terms of color, grain size, mineral composition, grain size, and particle packing.

Lithosequence. A set of soils with different properties due to differences in the parent materials.

Loam. A soil having a moderate amount of silt, sand, and clay. The U.S. Department of Agriculture textural class defines a loam as containing 7 to 27% clay, 28 to 50% silt, and less than 52% sand.

Loess. A geological deposit consisting of uniform, fine material, mostly silt, transported and deposited by wind.

Log K_{oc}. The soil/sediment partition or sorption coefficient is the ratio of adsorbed chemical per unit weight of organic carbon to the aqueous solute concentration. This value provides an indication of the tendency of a chemical to partition between particles containing organic carbon and water. Compounds that bind to organic carbon have characteristically low solubilities, whereas compounds with low tendencies to adsorb onto organic particles have high solubilities.

Log K_{ow}. The K_{ow} of a substance is the n-octanol/water partition coefficient and is the ratio of the solute concentration in the water-saturated n-octanol phase to the solute concentration in the n-octanol-saturated water phase.

Logarithm (log). A class of mathematical functions. If b and c are positive numbers and $b^x = c$, then x is the logarithm of c to the base b and is written as $\log_b c$. Many types of logarithms exist, including the natural logarithm, denoted as ln y or $\log_e y$ where e = 2.718, and common logarithms, which are represented as log y where $y = 10^x$.

Longitudinal dispersion (D_l). The spreading of a solute in groundwater in the direction of the bulk flow.

Longitudinal dunes. Long, narrow ridges of sand which are parallel to the direction of the prevailing wind.

Losing stream (influent stream). A stream or stretch of stream that contributes water to the saturation zone.

Low-temperature stripping. A soil remediation technology that utilizes a thermal processor to dry and heat soils up to 450°C. Volatile constituents are thereby stripped from the soil and can be recovered via condensation or adsorption on activated carbon or incinerated in an afterburner.

Lower explosive limit (LEL). The minimum concentration (volume % in air) of a flammable gas or vapor required for ignition or explosion to occur in the presence of an ignition source.

Lowest effects level (LEL). The lowest levels are the lowest acceptable effect levels for North American aquatic species reported in the toxicological literature. The acceptability of effect level data is based upon a critical review of a reference and comparison with established guidelines for toxicity testing.

Lyophilic. A substance that will readily go into colloidal suspension in a liquid.

Lyophobic. A substance in a colloidal state that has a tendency to repel liquids.

Lyotropic liquid crystal. A liquid crystal prepared by mixing two or more components, one of which is polar.

Lysimeter. (1) A device used to measure percolation and leaching losses from a column of soil under controlled conditions or to measure gains and losses. (2) A device used to collect soil pore water via suction in the unsaturated zone. Lysimeters are capable of retaining the accumulated water within a sampling vessel.

M

Macrolide. A large ring molecule with many functional groups bonded to it.

Macromolecule. A large molecule whose size may be in the lower colloidal range (0.5 to 1 micron). Such molecules are usually composed of either a sequence of chemically similar units linked together in chains of known length or an indefinite number of identical repeating units which may or may not be cross-linked. Examples include DNA, RNA, and protein.

Macropore. Soil pores that are significantly larger than the intergranular or interaggregate pores within the soil matrix. Macropores are large in relation to those pores in the surrounding soil and have the capability to conduct liquids faster than the surrounding micropores once flow is initiated.

Magnehelic gauge. A sensitive differential pressure or vacuum gauge manufactured by Dwyer Instrument Co. which uses a precision diaphragm to measure pressure differences. This gauge is manufactured in specific pressure or vacuum ranges such as 0 to 2 inches of water column. Magnehelic gauges are typically used to measure SVE system vacuums.

Magnesium acetate ($Mg(OOCCH_3)_2 \cdot 4H_2O$ or $Mg(OOCCH_3)_2$). A compound forming colorless crystals which are soluble and used in textile printing, in medicine as an antiseptic, and as a deodorant.

Magnesium arsenate ($Mg_3(AsO_4)_2 \cdot xH_2O$). A white, poisonous, water-insoluble powder which is used as an insecticide.

Magnesium borate ($3MgO \cdot B_2O_3$). A white crystal which is soluble in alcohol and acids.

Magnesium carbonate ($MgCO_3$). A water-insoluble, white powder which is used as refractory material.

Magnesium fluosilicate ($MgSiF_6 \cdot 6H_2O$; magnesium silicofluoride). Water-soluble, white crystals which are used in ceramics, in mothproofing, waterproofing, and as a concrete hardener.

Magnesium halide. A compound formed from the metal magnesium and any of the halide elements.

Magnesium hydride (MgH_2). A hydride compound formed from the metal magnesium.

Magnesium hydroxide ($Mg(OH)_2$). A white powder which is soluble in water; it is used as an intermediate in the extraction of magnesium metal, and as a reagent in the sulfite wood pulp process.

Magnesium oleate ($Mg(C_{18}H_{33}O_2)_2$). A water-insoluble material which is soluble in hydrocarbons, alcohol, and ether. It is used as a plasticizer lubricant and emulsifying agent and in varnish driers and drycleaning solutions.

Magnesium perchlorate ($Mg(ClO_4)_2 \cdot 6H_2O$). Deliquescent crystals which are soluble in water and alcohol and used as a drying agent for gases.

Magnesium peroxide (MgO_2). A powder which is soluble in dilute acids and is used as a bleaching and oxidizing agent.

Magnesium silicate ($3MgSiO_3 \cdot 5H_2O$). A white, water-insoluble powder which is used as a filler for rubber and in medicine.

Magnesium sulfite ($MgSo_3 \cdot 6H_2O$). A compound which is insoluble in alcohol and is used in medicine and paper pulp.

Malathion ($C_{10}H_{19}O_6PS_2$). A yellow liquid, slightly soluble in water, which is used as an insecticide.

Maleic hydrazide ($C_4H_2O_2$). A solid material slightly soluble in alcohol and water which is used as a weed killer and growth inhibitor.

Mandelic acid ($C_6H_5CH(OH)CN$). A white compound used in organic synthesis; also known as amygdalin acid, benzoglycolic acid, phenylglycolic acid, and α-phenylhydroxyacetic acid.

Maneb ($Mn[SSCH(CH_2)_2NHCSS]$). A generic term for manganese ethylene-1,2-bisdithiocarbamate.

Manganate. Salts that have manganese in the anion.

Manganous fluoride (MnF_2). A reddish powder that is soluble in acid; also known as manganese fluoride.

Manganous sulfate ($MnSO_4(4H_2O)$). A water-soluble compound used in medicine, textile printing, and ceramics; as a fungicide and fertilizer; and in paint manufacture.

Manganous sulfide (MnS). A powder used as a pigment and as an additive in making steel.

Manifest. An invoice for the shipment of hazardous waste. Regulations require every shipment of hazardous waste to have a manifest with copies submitted to state and regional offices.

Manifold. A pipe with several apertures for making multiple connections.

Mannich reaction. Reaction describing the condensation of a primary or secondary amine or ammonia with formaldehyde.

Mannitol hexanitrate ($C_6H_8(ONO_2)_6$). An explosive crystal which is soluble in alcohol, acetone, and ether and is insoluble in water. It is used in explosives and medicine.

Manometer. A device used to measure fluid or vapor pressures that consists of a tube filled with a liquid; the level of the liquid is determined by the fluid pressure, and the height of the liquid may be read from a scale.

Manure salts. Potash salts that have a high proportion of chloride and 20 to 30% potash.

Mask. A chrome-and-glass pattern for a layer of the wafer used in the photolithography process.

Mass. The amount of material substance present in a body, regardless of gravity.

Mass action. Law stating that the rate (speed, velocity) of a chemical reaction is directly related to the concentrations of the reactants, if the temperature remains constant.

Mass Action Law. The rate of a chemical reaction for a uniform system at constant temperature is proportional to the concentrations of the reacting substances.

Mass Conservation Law. In every chemical reaction or chemical change, the mass (weight) of the substances formed is precisely equal to that of the reacting substances, regardless of any alteration in their form or other properties.

Mass density. The mass of an object per unit volume.

Mass emission strategy (MES). Maximum concentration of a contaminant (in mg/kg on a wet-weight, edible-portion basis) that ensures that a consumer of the specified fish or wildlife species does not exceed the permissible intake level (PI) of the contaminant specified by the California Department of Health Services.

Mass formula. An equation giving the atomic mass of a nuclide as a function of its atomic number and mass number.

Mass movement. A generic term for various processes by which large masses of earth material are moved by gravity.

Mass number. The number of protons and neutrons in an atomic nuclide.

Mass shift. The portion of the isotope shift which results from the difference between the nuclear masses of different isotopes.

Mass spectrometry. A means of sorting out ions by measuring their masses according to their ratios of charge to mass. The procedure requires that a sample be vaporized and then ionized by electrons, either as the entire molecule or as constituent atoms. The charged particles are accelerated by an electric impulse and are passed through a magnetic field, which has the effect of changing their path from a straight line to a curve. Because some particles have greater mass than others, the effect is a separation of particles that is dependent on their ratio of mass to charge. Their impact on a photographic plate yields a spectrum from which the constituents can be determined.

Mass susceptibility. The magnetic susceptibility of a compound per gram; also known as specific susceptibility.

Mass-to-charge ratio. The measurement of a sample mass as a ratio to its ionic charge.

Material Safety Data Sheets (MSDS). Documents that describe the physical properties, exposure limits, and emergency procedures for specific hazardous chemicals.

Matric potential (capillary potential). Amount of work that must be done per unit quantity of pure water in order to transport reversibly and isothermally an infinitesimal quantity of water (identical in composition to the soil water) from a pool at the elevation and at the external gas pressure of the point under consideration, to the soil water.

Matrix. (1) The solid portion of a porous material. (2) Substances other than the analyte that are present in the sample or sample extract.

Matrix effects. The enhancement or suppression of minor element spectral lines from metallic oxides during emission spectroscopy.

Matrix smoothing. A technique used to smooth a grid of contoured data. This technique uses an averaging method or a weighted inverse distance method and a matrix or original grid values to create a new grid. Each original grid element is recalculated using the elements in the smoothing matrix and replaced. This technique has the effect of eliminating small-scale variability in the grid.

Matrix spectrophotometry. Spectrophotometric analysis in which the specimen is irradiated in sequence at more than one wavelength, with the visible spectrum evaluated for the energy leaving for each wavelength of irradiation.

Maximum contaminant level (MCL). The highest amount of a contaminant permitted in drinking water by the Safe Drinking Water Act. MCLs are the levels of contaminants in drinking water for which adverse health effects are not expected to occur over a lifetime of exposure. Although they are health-based criteria, MCLs also reflect the technological and economic feasibility of remediation of drinking water sources to MCL concentration.

Maximum contaminant level goals (MCLG). MCLGs are health-based goals for chemicals in drinking water. Concentrations proposed or promulgated do not consider technical or economical feasibility of implementation; therefore, they are generally much lower than an MCL (for example, MCLs for known human carcinogens are set at zero).

Maximum daily dose (MDD). The maximum daily dose an individual is likely to receive on any day during the period of his exposure.

Maximum possible concentration test. An extraction test method based on a shake test to determine contaminant solubility under saturated conditions. Distilled water is used as the leaching medium.

Maximum sustained yield. Maximum rate at which groundwater can be withdrawn perennially from a particular source.

Maximum water-holding capacity. The average moisture content of a disturbed soil sample, one centimeter high, that is at equilibrium with a water table at its lower surface.

Mean (arithmetic average). The sum of all measurements collected over a statistically significant period of time divided by the number of measurements.

Mean annual precipitation. Average annual amount of precipitation over a period of years.

Mean free path. A term used by physical chemists to denote the average distance traversed by an atom, molecule, or other particle without collision with another atom or molecule.

Mean particle density. The ratio of the density of a material to that of water at $4°C$ and at atmospheric pressure. The term is synonymous with density of solids and specific gravity. For mineral soils, the mean density of the particles ranges from 2.6 to 2.7 g/cm^3.

Mean sea level (MSL). The mean plane about which the tide oscillates. The average height of the sea for all stages of the tide.

Meander. One of a series of somewhat regular winding or looping bends in a river or stream.

Measurement bias. The consistent under- or overestimation of the true values in a population.

Measurement error. Error caused by the failure of the observed measurement to be the true value for the sampled unit.

Mecarbon ($C_{10}H_{20}O_5PS_2$). An oily liquid that is used as an insecticide and miticide for pests of rice, cabbage, and onions.

Mecarphon ($C_7H_{14}O_4PS_2$). A solid that is insoluble in water.

Mechanical aeration. A physical treatment process by which clean air is brought into contact with contaminated soil in order to transfer the volatile organics from the soil to the air stream. Subsequent treatment of the air stream is then applied.

Mechanical analysis. The procedure for determining the particle-size distribution of a soil sample.

Mechanical surging. A type of well development in which water is forced to flow into and out of a screen by forcing a plunger up and down in the well casing.

Mechanochemistry. The study of the conversion of mechanical energy into chemical energy in polymers.

Mecoprop ($C_{10}H_{11}ClO_3$). Crystalline compound used as an herbicide to control broadleaf weeds in cereals, turfs, and lawns.

Median. The middle value of a sample when the observations have been ordered from least to greatest. If there are an even number of measurements, the medium is the mean of the two central measurements

Median International Standard (MIS). The MIS is an indication of what other nations consider to be elevated contamination levels; it can only be used to provide general guidelines on other nations' findings. MISs apply to "flesh weight, edible portions" of freshwater fish and marine shellfish in parts per million (ppm) wet weight (ww), unless specifically noted otherwise.

Median sand. Sediment particles having diameters between 0.250 and 0.500 mm.

Medinoterb acetate ($C_{13}H_{16}N_2O_6$). A yellow, crystalline compound used as a pre-emergence herbicide for broadleaf weeds in cotton, sugarbeets, and leguminous crops and as a post-emergence herbicide for cereal crops.

Meinzer. A unit of hydraulic gradient expressed in gallons per day per square foot. This unit is equal to the flow rate in gallons per day through a cross-section of 1 foot square under a unit hydraulic gradient at 60°F.

Melanization. The darkening of a light-colored soil by the mixing of organic matter.

Membrane. An extremely thin, porous layer or film of material, either natural or synthetic; the thickness is approximately 100 angstrom units, with the diameter of the pores ranging from as little as 8 angstroms for natural cellular membranes up to 100 angstroms for manufactured membranes.

Membrane filter. A filter made of plastic with a known pore diameter.

Membrane separation. A physical treatment process in which solutes or contaminants from liquids are separated through the use of semi-permeable membranes.

Menadione ($C_{11}H_8O_2$). A yellow crystal which is soluble in alcohol, benzene, and vegetables oils and insoluble in water.

Menthene ($C_{10}H_{18}$). A colorless, water-insoluble, liquid hydrocarbon that is used in organic synthesis.

Menthyl ($C_{10}H_{19}$). A univalent radical that is formed from menthol by removal of the hydroxyl group.

Mephosfolan ($C_8H_{16}O_3PNS_2$). A yellow liquid which is used as an insecticide and miticide for agricultural crops.

Mercaptans (thioalcohols). A group of organosulfur compounds derived from hydrogen sulfide and found with other sulfur compounds in crude petroleum.

Mercaptide. A compound consisting of a metal and mercaptan.

Mercuric arsenate ($HgHAsO_4$). A poisonous powder used in anti-fouling and waterproof paints; also known as mercury arsenate.

Mercuric chloride ($HgCl_2$). A toxic compound that is soluble in alcohol or benzene. It is used in the manufacture of other mercuric compounds, as a fungicide, and in medicine and photography.

Mercuric oleate ($Hg(C_{18}H_{33}O_2)_2$). A poisonous, yellowish-to-red, liquid or solid mass which is insoluble in water and is used in medicine, anti-fouling paints, and as an antiseptic; also known as mercury oleate.

Mercuric sulfate ($HgSO_4$). A white, crystalline powder that is soluble in acid and is used as a catalyst and in galvanic batteries.

Mercuric thiocyanate ($Hg(SCN)_2$). A poisonous, white powder which is soluble in alcohol and slightly soluble in water and is used in photography; also known as mercuric sulfocyanide, mercury sulfocyanate, and mercury thiocyanate.

Mercury (Hg). A silvery metal, liquid at ordinary temperatures which is toxic by itself and in most compounds.

Mercury gauge. A gauge in which fluid pressure is measured by the height of a column of mercury which the fluid pressure will sustain.

Mercury naphthenate. A poisonous liquid that is soluble in mineral oils. It is used in gasoline anti-knock compounds and as a paint anti-mildew promoter.

Mesityl oxide ($(CH_3)_2C==CHCOCH_3$). An oily liquid that is used as a solvent for resins (particularly vinyl resins), many gums, and nitrocellulose in lacquers, paints, and varnishes.

Meso-. Prefix meaning intermediate or middle.

Mesomorphism. A state of matter that is as intermediate between a crystalline solid and a normal isotropic liquid.

Mesophilic digestion. Digestion by biological action at or below 45°C.

Metabolism. Term encompassing all of the diverse reactions by which a cell processes food material to obtain energy and the compounds from which new cell components are made.

Metal alkyl. A combination of an alkyl organic radical with a metal atom or atoms.

Metaldehyde ($(CH_3CHO)_n$). A white acetaldehyde-polymer that is soluble in organic solvents and insoluble in water.

Metallic bond. A strong electrostatic force existing between atoms in metallic crystals which binds the crystal together in a closely packed structure.

Metallic element. An element distinguished by its luster, electrical conductivity, malleability, and ability to form positive ions.

Metallic soap. A salt of stearic, oleic, palmitic, lauric, or eruic acid combined with a heavy metal. It is used as a drier in paints and inks and in fungicides, decolorizing varnish, and waterproofing.

Metallography. The study of the properties of metals and alloys as related to their physical structure, involving knowledge of crystal formation and structure, solid solution theory (phase rule), and microscopy at all levels.

Metallurgy. The body of scientific knowledge relating to the recovery, properties, and uses of metals.

Metamorphic facies. A group of metamorphic mineral assemblages that have reached chemical equilibrium during metamorphism within a prescribed range of temperature and pressure.

Metamorphic rocks. Any rock derived from pre-existing rocks by mineralogical, chemical, and/or structural changes, in response to marked changes in temperature, pressure, shearing stress, and chemical environment, generally at depth in the crust of the Earth.

Metamorphism. A change in the texture or mineralogical composition of a rock.

Methane (CH_4). The simplest hydrocarbon and the first member of the paraffin (alkane) series. Methane is a degradation product of carbonaceous materials and occurs in association with petroleum and coal, as well as in bogs and marshes.

Methanearsonic acid ($CH_3AsO(OH)_2$). A white solid which is soluble in water and is used as an herbicide for cotton crops and for noncrop areas.

Methanogenic. Referring to the formation of methane by certain anaerobic bacteria during the process of anaerobic fermentation.

Methazole ($C_9H_6Cl_2N_2O_3$). A solid that is used as an herbicide for control of weeds in crops.

Methidathion ($C_4H_{11}O_4N_2PS_3$). A crystalline compound that is used as an insecticide and miticide for pests on alfalfa, citrus, and cotton.

Methide. A binary compound consisting of methyl and a metal.

Methine group. A radical consisting of a single carbon and a single hydrogen.

Method detection limit (MDL). An analytical detection limit that is the minimum concentration of a substance that can be measured and reported with a 99% confidence level that the analyte concentration is greater than zero. The minimum detection limit is determined from analysis of a sample in a given matrix containing the analyte.

Methoxide. Compound formed from a metal and the methoxy radical.

2-Methoxyethanol ($CH_3OCH_2CH_2OH$). A poisonous liquid used as a solvent for cellulose acetate, leather, some synthetic resins, and alcohol-soluble dyes and in dyeing; also known as ethylene glycol monomethyl ether.

Methyl abletate ($C_{19}H_{29}COOCH_3$). A yellow liquid which is miscible with most organic solvents and is used as a solvent and plasticizer for lacquers, varnishes, and coatings.

Methyl acetate (CH_3CO_2CHI). A colorless liquid which is miscible with hydrocarbon solvents and is used as a solvent and extractant.

Methyl acrylate ($CH_2:CHCOOCH_3$). A volatile liquid which is slightly soluble in water and is used as a chemical intermediate and in making polymers.

Methyl allyl chloride ($CH_2:C(CH_3)CH_2Cl$). A volatile, flammable liquid which is used as an insecticide and fumigant and for chemical synthesis.

Methyl amyl alcohol ($(CH_3)_2CHCH_2CHOHCH_3$). A toxic flammable liquid which is miscible with water and most organic solvents and is used as a solvent and chemical intermediate; also known as methyl isobutyl carbinol (MIBC).

Methyl arachidate ($CH_3(CH_2)_{18}COOCH_3$). A wax-like solid which is soluble in alcohol and ether and is used in medical research and as a reference standard in gas chromatography.

Methyl bromide (CH_3Br). A toxic gas which forms a crystalline hydrate with cold water; it is used in the synthesis of organic compounds and as a fumigant.

2-Methyl-1-butanol ($C_5H_{12}O$). A liquid which is miscible with alcohol and ether and slightly soluble in water; it is used as a solvent, in organic synthesis, and as an additive in oils and paints.

Methyl butyl ketone ($CH_3COC_4H_9$). A liquid which is soluble in water, alcohol, and ether and is used as a solvent; also known as propylacetone.

Methyl butyrate ($CH_3CH_2CH_2COOCH_3$). A liquid which is used as a solvent for cellulosic materials.

Methyl caprate ($CH_3(CH_2)_4COOCH_3$). A combustible liquid which is soluble in alcohol and ether and is used in the manufacture of detergents, stabilizers, plasticizers, textiles, and lubricants.

Methyl chloroformate ($ClCOOCH_2$). A toxic, corrosive liquid used in military poison gas and insecticides.

Methyl cyclohexane ($CH_3C_5H_9CO$). A pale-yellow liquid used as a solvent and in lacquers.

Methy ethyl ketone ($CH_3COC_2H_5$). A water-soluble liquid which is miscible in oil.

Methyl glucoside ($C_7H_{14}O_6$). Crystals used to make resins, drying oils, plasticizers, and surfactants.

Methyl hydroxystearate ($C_{19}H_{38}O_3$). A waxy material which is slightly soluble in organic solvents and insoluble in water; it is used in cosmetics, inks, and adhesives.

Methyl isobutyl ketone ($(CH_3)_2CHCH_2COCH_3$; MIK). Clear liquid with a characteristic odor. Short-term exposure to MIK can cause irritation of the conjunctiva and mucous membranes of the nose and throat and produce eye and throat symptoms.

Methyl lactate ($CH_3CHCHCOOCH_3$). A liquid which is miscible with water and most organic liquids and is used as a solvent for lacquers, stains, and cellulosic materials.

Methyl laurate ($CH_3(CH_2)_{10}COOCH_3$). A water-insoluble liquid used as a chemical intermediate to make rust removers and for leather treatment.

Methyl linoleate ($C_{19}H_{34}O_2$). A colorless liquid which is soluble in alcohol and ether and is used in the manufacture of detergents, emulsifiers, lubricants, and textiles and in medical research.

Methyl methacrylate ($CH_2C(CH_3)COOCH_3$). A flammable liquid which is soluble in most organic solvents but insoluble in water; it is used as a monomer for polymethacrylate resins.

Methyl myristate ($CH_3(CH_2)_{12}COOCH$). A colorless liquid used in the manufacture of detergents, plasticizers, resins, textiles, animal feeds, and as a flavoring agent.

Methyl-*n*-amyl ketone ($CH_3(CH_2)_4CHOHCH_3$). A stable liquid which is miscible with organic lacquer solvents and is used as reaction medium and solvent for nitrocellulose lacquers.

Methyl oleate ($C_{17}H_{33}COOCH_3$). An amber liquid which is soluble in organic liquids, mineral spirits, and vegetable oil and is insoluble in water; it is used as a plasticizer and softener.

Methyl parathion ($C_8H_{10}NO_5PS$). A dark-brown liquid which is slight solubility in water and is used as an insecticide to control boll weevils, leafhoppers, cutworms, and rice bugs.

Methyl propionate ($CH_3CH_2COOCH_3$). A flammable liquid that is soluble in most organic solvents; it is used as a solvent for cellulose nitrate; in lacquers, varnishes, and paints; and for flavoring.

Methyl ricinoleate ($C_{19}H_{36}O_3$). A low-viscosity fluid which is used as a wetting agent, cutting oil additive, lubricant, and plasticizer.

Methyl silicone ($[(CH_3)_2SiO]_x$, $[C(CH_3)_2Si_2O_3]_y$, etc.). A silicone with properties of oil, resin, or rubber, depending on molecular size and arrangement.

Methyl stearate ($C_{17}H_{35}COOCH_3$). Crystals which are soluble in alcohol and ether, insoluble in water, and used as an intermediate for stearic acid manufacture.

Methyl styrene ($C_6H_5C(CH_3):CH_2$). A toxic liquid used to produce polystyrene resins.

Methyl tertiary butyl ether (MTBE). A chemical used as an octane enhancer for gasoline.

Methylal ($CH_3OCH_2OCH_3$). A liquid which is soluble in ether, hydrocarbons, alcohol, and water. It is used as a solvent and chemical intermediate and in perfumes, adhesives, and coatings.

Methylamine (CH_3NH_2). A gas used to prepare dyes and as a chemical intermediate; also known as aminoethane and monomethylamine.

N-Methylaniline ($C_6H_5NH(CH_3)$). An oily liquid which is soluble in water and organic solvents and is used as an acid acceptor, solvent, and chemical intermediate.

Methylarsinic sulfide (CH_3AsS). A colorless compound which is insoluble in water and is used as a fungicide in treating cotton seeds; also known as rhizoctol.

Methylbutynol ($HC; CCOH(CH_3)_2$). A water-miscible liquid that is soluble in most organic solvents. It is used as a stabilizer for chlorinated organic compounds, as a solvent, and as chemical intermediate.

Methylcellulose. A grayish-white powder used in water-based paints and ceramic glazes, for leather tanning, and as a thickening and sizing agent, adhesive, and food additive.

Methylene bromide (CH_2Br_2). A liquid which is miscible with organic solvents and slightly soluble in water; it is used as a solvent and chemical intermediate.

Methylene chloride (CH_2Cl_2). A colorless liquid which has an odor like chloroform and is an anesthetic; short-term exposure to methylene chloride can cause mental confusion, light-headedness, nausea, vomiting, and headache. High vapor concentrations may also cause irritation of the eyes and respiratory tract.

Methylethylcellulose. A white to cream-colored, fibrous solid or powder used as an emulsifier and foaming agent.

2-Methylfuran ($C_4H_3OCH_3$). A colorless liquid used as a chemical intermediate.

Methylisothiocyanate (C_2H_3NS). A crystalline compound that is soluble in alcohol and ether. It is used as a pesticide and in amino acid sequence analysis.

Methylmercury nitrile (CH_3HgCN). A crystalline solid used as a fungicide to treat seeds of cereals, flax, and cotton.

Methylnaphthalene ($C_{10}H_7CH_3$). A solid used in insecticides and organic synthesis.

Methylol urea ($H_2NCONHCH_2OH$). Water-soluble crystals used to treat textiles and wood and in the manufacture of resins and adhesives.

Methylpentanoic acid ($(CH_3)_2CH(CH_2)_2COOH$). A colorless liquid which is soluble in alcohol, benzene, and acetone; it is used for plasticizers, vinyl stabilizers, and metallic salts.

Methylphosphoric acid ($(CH_3)_2CH(CH_2)_2COOH$). A liquid used for textile- and paper-processing compounds, as a rust remover, and in soldering flux.

Metobromuron ($C_9H_{11}BrN_2O_2$). A crystalline compound used as a pre-emergence herbicide to control weeds in potatoes.

Metribuzin ($C_8H_{14}N_4OS$). A crystalline solid used as a pre-emergence herbicide for soybeans and pre- and post-emergence treatment for potatoes.

Micelle. A unit or structure built up from complex molecules in colloids.

Microaerophilic. Obligate aerobes that function best under conditions of low oxygen concentration.

Microanalysis. Analysis of samples weighing from 1/10 to 10 milligrams.

Microbar. A unit of pressure equal to 1 dyne per square centimeter.

Microbes. Microscopic organisms.

Microbial activity. Chemical changes resulting from the metabolism of living organisms.

Microbiota. Microscopic plants and animals.

Microchemistry. The study of chemical analysis of material on a small scale so that specialized instruments such as the microscope are needed; the material analyzed may be on the scale of 1 microgram.

Microcosm. A diminutive, representative system analogous to a larger system in composition, development, or configuration. As used in biodegradation treatability studies, microcosms are typically constructed in glass bottles or jars.

Microcrystalline. A term used to describe materials that can be prepared in a form in which the crystals are much smaller than in the natural product.

Microcurie (μCi). Unit of radioactivity equal to 10^{-6} curie.

Microdensitometer. A densitometer used in spectroscopy to detect spectrum lines too faint to be seen by the human eye.

Microelectrophoresis. The direct microscopic observation and measurement of the velocity of migration of ions or other charged bodies through a solution toward oppositely charged electrodes.

Microfractures. Microscopic fractures in rocks on the order of a millimeter or less in length.

Microgram (μg). A unit of weight equal to one thousandth of a milligram; one millionth of a gram; one billionth of a kilogram.

Microorganism. A minute organism, either plant or animal, invisible or barely visible to the naked eye.

Microradiography. A technique used to study surfaces of solids by monochromatic-radiation contrast effects shown via projection or enlargement of a contact radiograph.

Microwave spectroscopy. A type of absorption spectroscopy utilizing the portion of the electromagnetic spectrum lying between the far and radio frequencies (i.e., wavelengths from 1 to 1000 millimeters). Substances analyzed must be in the gas phase.

Mil. One mil equals 1/1000 of an inch.

Millibar. A unit of pressure equal to 1000 dynes per square centimeter, or 1/1000 of a bar at atmospheric pressure.

Millicurie (mCi). A unit of radioactivity equal to 10^{-3} curie.

Millidarcy. One-thousandth Darcy; equal to 0.01824 gallons per day per square foot at 60°F.

Milligrams per liter (mg/L). A weight-to-volume relationship used to describe a weight-to-liquid ratio. Milligram per liter is equivalent to grams per cubic meter.

Milliliter (mL). One thousandth of a liter; equal to the volume occupied by 1 gram of water at 4°C (40°F) and 1 atmosphere pressure.

Millimicron. Unit of length equal to one thousandth of a micron.

Mineral horizon. A soil horizon that contains less than 12% organic carbon if the mineral fraction contains no clay, or less than 18% organic carbon if the mineral fraction contains 60% or more clay.

Mineral spirits (petroleum spirits, white spirits, or Stoddard solvents). A hydrocarbon derived from the light distillate fractions during the crude-oil refining process. It is composed of the C_6 to C_{11} compounds, with the majority of the relative mass composed of C_9 to C_{11}. Mineral spirits are composed of these general classes of compounds: 50% paraffins, 40% cycloparaffins, and 10% aromatics.

Mineralization. Removal of oxide solids through the decomposition of organic matter.

Mineralogy. The study of minerals, including their formation, occurrence, properties, composition, and classification.

Misch metal. A commercially made alloy or mixture of several rare-earth metals, of which cerium constitutes about 50%.

Miscible. A term describing the extent to which liquids or gases can be mixed or blended.

Miscibility. Tendency or capacity of two or more liquids to form a uniform blend.

Mississippian. A geologic epoch between 360 and 320 million years ago.

Mists. Suspended liquid droplets (<2 micrometers) generated by condensation from the gaseous to the liquid state or by breaking up a liquid into a dispersed state, such as by splashing, foaming, or atomizing. Mist is formed when a finely divided liquid is suspended in air.

Mitigation. The reduction or alleviation of a problem — for example, the process of cleaning up a contaminated site in order to return it to an environmentally acceptable state.

Mixed aniline point. The minimum temperature at which a mixture of aniline, heptane, and hydrocarbon will form a solution.

Mixed potential. The electrode potential of a material while more than one electrochemical reaction is occurring simultaneously.

Mixing chamber. A chamber used to facilitate the mixing of chemicals with two or more liquids of different characteristics.

Mixture. A mechanical blend of two or more substances in any proportions; the components may or may not be uniformly dispersed. The substances are not united chemically and can be separated by a number of means, as by filtration, distillation, and precipitation.

Mobile treatment. Modular equipment and the corresponding processes that can be brought to a hazardous waste site and transported to a number of sites. The equipment is generally smaller than conventional equipment used in permanent structures.

Mode. The value that occurs most frequently in a group of numbers.

Modulus of elasticity. The ratio of the change in stress to the change in strain.

Mohr circle. A graphical representation of the state of stress (normal and shear) on a particular plane inclined at an angle to the major principal stress.

Mohr titration. Titration of a fluid with silver nitrate to determine the concentration of chlorides in a solution.

Moisture density. Mass of water per unit volume of space occupied by soil, air, and water.

Moisture tension. A numerical measure of the energy with which water is held in the soil.

Molal quantity. Number of moles present, expressed with weight in pounds, grams, or other such units, numerically equal to the molecular weight; for example, the pound-mole or gram-mole.

Molal solution. Concentration of a solution expressed in moles of solute divided by 1000 grams of solvent.

Molality. Number of moles of solute that are dissolved in a mass (1 kilogram) of solution.

Molar conductivity. Ratio of the conductivity of a electrolytic solution to the concentration of electrolyte in moles per unit volume.

Molar gas constant (R). A fundamental physical constant equal to 8.3145 J mol^{-1} K^{-1}.

Molarity. The measure of the number of gram-molecular weights of a compound present (dissolved) in 1 liter of solution.

Mole. The weight of a substance expressed in grams. A mole is an amount containing the same number of units as there are atoms in 12 grams of carbon-12, (i.e., 6.023×10^{23}). In general terms, a mole is the Avogadro's number of any chemical unit.

Mole fraction. The ratio of moles of a component in a mixture to the total number of moles in the mixture.

Molecular adhesion. An intermolecular phenomenon in which solids or liquids adhere to each other.

Molecular attraction. A force that pulls molecules toward each other.

Molecular biology. That portion of biochemistry devoted largely to the study of genetic mechanisms at the molecular level.

Molecular diffusion. Dispersion of a chemical caused by the kinetic activity of the ionic or molecular constituents.

Molecular distillation. A process by which substances are distilled in high vacuum at the lowest possible temperature and with the least damage to their composition.

Molecular gas. A gas composed of a single species, such as oxygen, chlorine, or neon.

Molecular sieve. A term used to describe the function of zeolite materials, which are clay-like in chemical nature (aluminosilicate compounds) and from which all water can be removed without alteration of their molecular structure. As a result of this, the material becomes microporous to such an extent that about half of its volume is occupied by very small holes or channels (cages). The material readily adsorbs molecules of other substances in much the same manner as activated carbon.

Molecular still. Apparatus used to conduct molecular distillation.

Molecular vibration. The theory that all atoms within a molecule are in continuous motion.

Molecular volume. Volume occupied by 1 mole (gram-molecular weight) of an element or compound.

Molecular weight. The molecular weight is the sum of the atomic weights of all of the atoms in a molecule, expressed in grams per mole.

Molecule. The smallest part of a compound or element that exists independently and which possesses the chemical properties of the element or compound.

Mollic epipedon. A surface soil horizon that contains 1% or more organic matter with a base saturation over 50%.

Molybdenum dioxide (MoO_2). A lead-gray powder used in pigment for textiles.

Molybdenum disulfide (MoS_2). A black powder which is insoluble in water and is used as a dry lubricant and additive for greases and oils.

Molybdenum trioxide (MoO_3). A white solid that is soluble in concentrated mixtures of nitric sulfuric acids and nitric and hydrochloric acids. It is used as a corrosion inhibitor, in enamels and ceramic glazes, in medicine and agriculture, and as a catalyst in the petroleum industry.

Monel®. A corrosion-resistant alloy containing 67% nickel and 30% copper; the remaining 3% is either aluminum or silicone.

Monitoring. The periodic or continuous review of a program, facility, or environment.

Monitoring well. A well installed for routine observation of groundwater levels or systematic collection of water samples to be analyzed for chemical pollution.

Mono. (1) Prefix derived from the Greek word for "one"; (2) prefix for chemical compounds to show a single radical.

Monoacetate. A compound such as a salt or ester that contains one acetate group.

Monoaromatic. Aromatic hydrocarbons containing a single benzene ring.

Monochromator. (1) Device for isolating monochromatic or narrow bands of radiant energy from the source; (2) a spectrograph in which the detector is replaced by a second slit, placed in the focal plane, to isolate a particular narrow band of wavelengths for refocusing on a detector or experimental object.

Monocyclic aromatic hydrocarbons (MAHs). The class of aromatic hydrocarbons which contain a single benzene ring.

Monod kinetic rate equation. A rate equation used to describe the growth rate of a single species or mixed species population of microorganisms. It is expressed as:

$$V = V_{max} \; C/(K + C)$$

where

$$
\begin{aligned}
V &= \text{specific growth rate of the microorganisms.} \\
V_{max} &= \text{maximum growth rate of the microorganisms.} \\
C &= \text{concentration of organic chemical.} \\
K &= \text{organic chemical concentration supporting a growth rate} \\
&\quad \text{equal to one half of the maximum } (V_{max}/2).
\end{aligned}
$$

Monodisperse. For soils, a soil whose grain diameter is uniform; one that includes various grain sizes is termed polydisperse.

Monomer. (1) A chemical compound consisting of single compounds; (2) a simple molecule capable of combining with a number of like or unlike molecules to form a polymer.

Monooxygenase. An enzyme that catalyzes reactions in which one atom of O_2 appears in the product and the other in H_2O.

Monovalent. A radical or atom whose valency is 1.

Monoxide. A compound that contains a single oxygen atom.

Montmorillonite. An aluminosilicate clay mineral with 2:1 expandable layer structure (two silicone tetrahedral sheets enclosing an aluminum octahedral sheet). Considerable expansion may be caused by water moving between the silica layers of contiguous units.

Mor. The surficial forest soil horizon that consists of acid litter and humus.

Moraine. Mound, ridge, or other accumulation of unsorted, unstratified glacial drift, deposited chiefly by the direct action of glacier ice.

Moraine. A mass of loose material deposited by a glacier in the process of melting.

Mordant (dye mordant). An agent that fixes dyes to tissues, cells, textiles, and other materials by combining with the dye to form an insoluble compound; a term used in dyeing technology to indicate a substance that has the property of fixing or binding dye molecules to textile fibers.

Morpholine (C_4H_8ONH). A liquid used as a solvent and rubber accelerator.

Mottled. A soil horizon that is irregularly marked with spots of color; mottling is caused by imperfect soil drainage and/or different minerals.

Mounding. Phenomenon usually created by the recharge of groundwater from a manmade structure into a permeable geologic material. Associated groundwater flow will be away from the structure in a radial fashion.

Moving average. Procedure used to smooth time series data by calculating a weighted moving average of points in a time series.

Mucic acid ($HOOC(CHOH)_4COOH$). A white material used as a metal ion sequestrant and to retard concrete hardening.

Muck soil. Organic soil derived from the decomposition of peat.

Mud. A sticky mixture of water and finely divided particles of a solid such as dirt which exhibits little or no plasticity.

Mudflow. A well-mixed mass of water and alluvium with high viscosity and low fluidity which moves at a slow rate, usually piling up and spreading over the ground like a sheet of wet concrete.

Muffled soil. A soil that has been baked in a 400°F muffle furnace for 4 hours to remove volatile organics.

Mull. A forest soil consisting of mineral matter and amorphous humus.

Mull technique. A method for obtaining infrared spectra of materials in the solid state; the material to be scanned is first pulverized, then mixed with mineral oil.

Multiphase flow. The movement of one or more immiscible fluid phases through porous media.

Multiple comparison procedure. Statistical procedure that makes a large number of decisions or comparisons from one set of data.

Mutagenicity. The ability of a substance to produce a detectable and heritable change in genetic material, which may be transmitted to the progeny of affected individuals through germ cells (germinal mutation) or from one cell generation to another within the individual (somatic mutation).

Myristyl alcohol ($C_{14}H_{29}OH$). A liquid used as a chemical intermediate, plasticizer, and perfume fixative.

N

N electron. An electron in the fourth (N) shell of electrons surrounding the atomic nucleus, having the principal quantum number 4.

Nano. A prefix meaning one-billionth part.

Nanocurie (nCi). A unit of radioactivity equal to 10^{-9} curie.

Nanogram. One thousandth of one millionth of a gram, or 10^{-9} gram.

Naphtha. Any of several liquid mixtures of hydrocarbons of specific boiling and distillation ranges derived from either petroleum or coal tar. As a group, they are highly volatile and flammable; the grades derived from coal are toxic.

Naphthacene ($C_{18}H_{12}$). A hydrocarbon molecule with four benzene rings fused together.

Naphthalene ($C_{10}H_8$; naphthalin; tar camphor). Volatile crystals with a coal tar aroma used for moth repellents, fungicides, lubricants, and resins and as a solvent.

Naphthenate. An oil-soluble metallic salt of naphthionic acid, the metals used being copper, cobalt, lead, manganese, and sodium. The products are rather thick, heavy liquids or pastes; they have strong fungicidal properties and are used to impregnate fabrics exposed to tropical insects and in anti-fouling paints.

Naphthene. Any of the cycloparaffin derivatives of cyclopentane (C_5H_{10}) or cyclohexane (C_6H_{12}) which are found in crude petroleum.

Naphthenic acid. A derivative of cyclopentane, cyclohexane, cycloheptane, or other naphthenic homologues derived from petroleum with molecular weights of 180 to 350. It is soluble in organic solvents and hydrocarbons and slightly soluble in water. It is used as a paint drier and a wood preservative and in metals production.

National Ambient Air Quality Standards (NAAQS). NAAQS are established under the Clean Air Act and represent health-based standards for chemicals in air.

National Pollutant Discharge Elimination System (NPDES). The national program for issuing, modifying, revoking and reissuing, terminating, monitoring, and enforcing permits and imposing and enforcing pretreatment requirements under sections 307, 318, 402, and 405 of the Clean Water Act. The term includes approved state programs.

Natric horizon. A subsurface soil that has more than 15% of its cation exchange capacity saturated with sodium.

Natural degradation. The degradation of organic matter via chemical reactions such as hydrolysis, oxidation, and reduction and microbiological reactions.

Natural logarithm. Logarithms to the base e. The number e is an irrational number that can only be approximated:

$$e = 2.7182818284\ldots$$

Nautical mile. Unit of distance used in ocean navigation; the international nautical mile is 1852 meters (6070.10 feet).

Neat cement. A slurry of water and cement of a consistency that can be forced through a pipe and placed as required.

Neburon ($C_{12}H_{16}C_{l2}N_2O$). A white, crystalline compound with a melting point of 102 to 103°C which is used as an herbicide to control weeds in nursery ornamentals, dichondras, and wheat.

Negative catalysis. A catalytic reaction such that the reaction is slowed down by the presence of the catalyst.

Negative ion. An atom or group of atoms which, by gain of one or more electrons, has a negative electric charge.

Nematicides. All substances or mixtures of substances used to control or destroy nematodes.

Neo-, ne-. Prefix indicating hydrocarbons having a carbon bonded directly to at least four other carbon atoms.

Neodymium oxide (Nd_2O_3). A blue-gray powder which is soluble in acids and is used to color glass and in ceramic capacitors.

Neohexane (C_6H_{14}). A volatile, flammable liquid which is used as a high-octane component of motor and aviation gasolines.

Neopentane (C_5H_{12}). A liquid which is soluble in alcohol and insoluble in water; it is a hydrocarbon found as a minor component of natural gasoline.

Nephelometry. A technique used to measure the turbidity of a solution. The light reflected at right angles when passing through the solution is recorded as a measurement of the dispersed particles in the solution.

Nessler tubes. Matched cylinders with strain-free, clear-glass bottoms for comparing color density or opacity.

Net peak flow. The total flow at a peak, minus the corresponding base flow.

Neutral soil. A soil that is neither acid nor alkaline. In practice, it is a soil with a pH between 6.6 and 7.3.

Neutralization. A chemical treatment process in which the interaction of an acid or base with another solution results in solution or mixture with a pH between 5 and 9.

Neutralization number. The milligrams of potassium hydroxide required to neutralize the acid in 1 gram of oil; it is used as an indication of oil acidity.

Neutralize. To make a solution neutral by adding a base to an acidic solution or an acid to a basic solution.

Neutron. An elementary particle that has no electric charge.

Neutron number. The number of neutrons in the nucleus of an atom.

Newton (N). A unit of force equal to the force required to accelerate a 1 kilogram mass at 1 meter per second. One Newton equals 0.2248 pounds of force.

Newtonian viscous fluids. Fluids in which the magnitude of shear stress is approximately linearly related to the rate of deformation.

Nickel (Ni). A hard, ductile, malleable, silver-white metallic element of the iron-cobalt group.

Nickel arsenate ($Ni_3(AsO_4)_2(H_2O)$). A poisonous yellow-green powder used as a fat-hardening catalyst in soapmaking.

Nickel carbonate ($NiCO_3$). Crystals that decompose upon heating and are soluble in acids, insoluble in water, and used in electroplating.

Nickel cyanide ($Ni(CN)_2(4H_2O)$). A poisonous powder used for electroplating.

Nigre. In soap technology, the brown-to-black liquid remaining after the soap has been separated from the other products of the saponification reaction by the salting-out process. It contains glycerol, sodium chloride, and various impurities.

Nitralin ($C_{13}H_{19}N_3O_6S$). A light yellow to orange crystalline compound used as a pre-emergence herbicide for weed control for cotton, food crops, and ornamentals.

Nitrate. A compound characterized by the presence of one or more NO_3 groups.

Nitric acid (HNO_3). A strong oxidant which is a fire hazard. It is used for chemical synthesis, explosives, fertilizer manufacture, etching, engraving, and ore flotation.

Nitride. A compound of nitrogen and a metal.

Nitrification. Conversion of nitrogenous matter into nitrate by bacteria.

Nitrile. A term used to designate an organic compound in which the univalent cyanogen group (CN) is present; the parent substance is hydrocyanic acid, HCN.

Nitrite. A organic or inorganic compound containing the radical NO_2^-.

Nitroalkane. A compound in which the hydrogen atom of an alkane molecule has been replaced by a nitro group.

Nitrobacteria. Bacteria that oxidize nitrite to nitrate.

Nitrocellulose. Compounds created by reacting cellulose with nitric or a mixed acid.

Nitroethane ($CH_3CH_2NO_2$). A colorless liquid used as a solvent for cellulosics, resins, waxes, fats, and dyestuffs and as a chemical intermediate.

Nitrogen dioxide (NO_2). A poisonous reddish-brown gas which, when cooled, changes to a light-yellow liquid (nitrogen tetroxide; N_2O_4), becoming a crystalline solid without color at $-15.3°F$.

Nitrogen fixation. The utilization of nitrogen from the air by plants to form critically important compounds. Nitrogen conversion is performed by certain types of bacteria which live symbiotically on the root hairs of plants, thus enabling them to synthesize nitrogenous compounds (proteins).

Nitrogen solution. A mixture used to neutralize super-phosphate in fertilizer manufacturing. It consists of 60% ammonium nitrate, with the balance being an aqua ammonia solution.

Nitrometer. A glass apparatus used to collect and measure nitrogen and other gases evolved by a chemical reaction.

Nitromethane (CH_3NO_2). A liquid nitroparaffin compound used as a monopropellant for rockets, in chemical synthesis, and as an industrial solvent for cellulosics, resins, waxes, fats, and dyestuffs.

Nitroso-. Prefix indicating the presence of an NO group in an organic compound.

Nitrosomonas. A group of bacteria capable of converting ammonia to nitrite under aerobic conditions.

Nivation. The erosion behind and peripheral to a snow bank, caused by frost action, mass movement, transport by melt water, or other related processes.

Nodulizing. A method of ore preparation in which ore fines are heated in an oil- or gas-fired rotary kiln.

Nomograph. A diagram for the graphical solution of problems that involve formulas in two or more variables.

Non-aqueous phase liquid (NAPL). Contaminants that remain as the original bulk liquid in the subsurface.

Noncarcinogenic polycyclic aromatic hyrocarbons (PAH). These include the following compounds: naphthalene, acenaphthylene, acenaphthene, fluorene, phenanthrene, anthracene, fluoranthene, and pyrene.

Nonconformity. A time break in a sequence of depositions where the sedimentary deposits rest on older igneous or metamorphic rocks.

Nonindurated deposits. Depositions composed of particles of gravel, sand, silt, or clay which are not bound or hardened by mineral cement, pressure, or the thermal alteration of the grains.

Nonionic surfactant. A general family of surfactants, so called because in solution the entire molecule remains associated.

Nonlinear spectroscopy. The study of energy levels not normally accessible with optical spectroscopy through the use of nonlinear effects such as multiphoton absorption and ionization.

Nonoic acid ($C_8H_{17}COOH$). A family of acids which are mixed isomers produced in the Fischer-Tropsch process.

Nonparametric method. A technique that does not depend for its validity on the data drawn from a specific distribution, such as the normal or log normal distribution.

Nonplastic. A soil morphology term that qualitatively describes the soil plasticity. It denotes a soil that can be rolled into a wire between the hands.

Nonpoint source. Sources of water pollution that are not associated with point sources, such as agricultural fertilizer runoff or sediment from construction. Examples include: (1) agriculturally related nonpoint sources of pollution, including runoff from manure disposal areas and from land used for livestock and crop production; (2) silvicultural-related nonpoint sources of pollution; (3) mine-related sources of pollution, including new, current, and abandoned surface and underground mine runoff; (4) construction activity-related sources of pollution; (5) sources of pollution from disposal on land, in wells, or in subsurface excavations that affect ground surface water quality; (6) saltwater intrusion into rivers, lakes, estuaries, and groundwater resulting from reduction of freshwater flow from any cause, including irrigation, obstruction, ground-water extraction, and diversion; and (7) sources of pollution related to hydrologic modifications, including those caused by changes in the movement, flow, or circulation of any navigable water or groundwater due to construction and operation of dams, levees, channels, or flow diversion facilities.

Nonpolar. Pertaining to an element or compound which has no permanent electric dipole moment.

Normal compound. An organic compound whose structure is of the straight- or open-chain type, containing no branching or subsidiary groups.

Normal distribution. The symmetrical (bell-shaped) distribution of data characterized by the mean and variance of the data.

Normal salt. A salt in which all of the acid hydrogen atoms are replaced by a metal, or the hydroxide radicals of a base are replaced by an acid radical.

Normal solution. A solution that contains one chemical equivalent weight of a substance per liter of solution.

Normal state. A term sometimes used for ground state.

Normality (N). The measure of the number of gram-equivalent weights of a compound per liter of solution.

Normalize. A term used by metallurgists to refer to a process of tempering steels and other alloys by heating them to a predetermined temperature and cooling at a controlled rate to relieve internal stresses and improve strength and stability.

Nuclear absorption. The absorption of energy by the nucleus of an atom.

Nuclear binding energy. The energy required to separate an atom into its constituent protons, neutrons, and electrons.

Nuclear capture. A process in which a particle — such as a neutron, proton, electron, muon, or alpha particle — combines with a nucleus.

Nuclear chemistry. The study of transformations in the nucleus of atoms.

Nuclear magnetic resonance (NMR). Analytical technique used to measure the nuclei of material when placed in a magnetic field that is subjected to an alternating magnetic field. This method is useful for distinguishing between nuclear particles present in a sample.

Nuclear magnetic resonance spectrometer. A spectrometer in which nuclear magnetic resonance is used for the analysis of protons and nuclei. It is used to study changes in chemical and physical quantities over wide frequency ranges.

Nucleation. The process by which the phase change of a substance to a more condensed state is initiated at a certain loci within the less condensed state of the substance.

Nucleophile. A reacting specie that brings an electron pair.

Nucleus. (1) The central portion of a cell, separated by a membrane from the cytoplasm and containing the chromosomes; (2) small, positively charged core of an atom.

Nuclide. A term used by nuclear physicists to refer to any isotopic form of any element.

Nunatak. An isolated hill or peak that projects through the surface of a glacier.

Nutrients. Major elements (e.g., nitrogen and phosphorus) and trace elements (including sulfur, potassium, calcium, and magnesium) that are essential for the growth of organisms.

O

Obligate aerobes. Organisms requiring the presence of molecular oxygen (O_2) for their metabolism.

Obligate anaerobes. Organisms for which the presence of molecular oxygen is toxic. These organisms derive the oxygen needed for cell synthesis from chemical compounds.

Observation well. A well drilled in a selected location for the purpose of observing parameters such as water levels and pressure changes.

Occlude. To cause to become obstructed or closed and thus prevent passage either in or out.

Octadecene ($C_{18}H_{36}$). A colorless liquid which is soluble in alcohol, acetone, ether, and petroleum and insoluble in water.

Octadecenyl aldehyde ($C_{17}H_{35}CHO$). A flammable liquid used in the manufacture of vulcanization accelerators, rubber antioxidants, and pesticides.

Octane. A paraffinic hydrocarbon (C_8H_{18}) obtained from petroleum. Its isomer, isooctane, is a branched-chain structure which is the basis of high-octane gasoline. Both have good solvent properties and are used in the synthesis of organic chemicals.

n-**Octane (C_8H_{18})**. A colorless liquid which is soluble in alcohol, acetone, and ether and insoluble in water. It is used as a solvent chemical intermediate.

Octane number. An arbitrary value denoting the anti-knock rating of a gasoline. It is the proportion of the branched-chain hydrocarbon isooctane that is contained in a test mixture of heptane and isooctane.

Octanol-water partition coefficient (K_{ow}). A coefficient representing the ratio of solubility of a compound in octanol to its solubility in water. As the octanol-water partition coefficient increases, the water solubility of the chemical decreases. The greater the value of the octanol partition coefficient, the more the chemical will be adsorbed to soil. The K_{ow} is indirectly related to water solubility and directly related to soil/sediment adsorption coefficients and bioconcentration factors for aquatic biota. Chemicals with K_{ow} values less than 10 (log K_{ow} less than 1) generally have high water solubilities, low soil/sediment adsorption coefficients, and low bioconcentration factors. It provides a measurement of the hydrophobic nature of a chemical and can be measured in the laboratory for most chemicals.

Octant search method. A term associated with data gridding in which the original data points used to estimate a grid element are found by dividing the area around a grid element into eight equal octants.

Octene ($CH_3(CH_2)_5CHCH_2$). A flammable liquid which is used as a plasticizer and in the synthesis of organic compounds.

Octyl phenol ($C_8H_{17}C_6H_4OH$). White flakes that are soluble in organic solvents and insoluble in water.

Odor threshold. The minimum concentration of a substance at which a majority of test subjects can detect and identify the characteristic odor of the substance.

Off-gas treatment system. Refers to the unit operations used to treat (i.e., condense, collect, or destroy) contaminants in the purge gas from the thermal desorber.

Offshore bar. A beach, essentially parallel to the shoreline, formed some distance from the shoreline.

Offshore terrace. A sand deposit that is built out into deep water by the combined action of waves and currents.

Ohm. The resistance through which a difference in 1 volt will produce a current of 1 ampere.

Ohm's Law. A law that states that the current, or flow rate, of electricity is proportional to the electrical potential gradient.

-Oic. A suffix indicating the presence of a –COOH group.

Oil and grease. Hydrocarbons, fatty acids, soaps, fats, waxes, oils, and other material which are extracted by solvents from an acidified sample and are not volatilized during the test.

Oil blue. A violet-blue copper sulfide pigment used in varnishes.

Oil gas. A gas rich in hydrogen, methane, ethane, and other light hydrocarbon gases (heating value of about 1000 Btu/ft^3) made by pyrolysis of petroleum oils.

Oils. A nonspecific term applied to several groups of organic mixtures which include petroleum oils and lubricants.

-Ol. A chemical suffix for an –OH group in organic compounds.

Olefin. A major group of aliphatic hydrocarbons, both straight- and branched-chain, characterized by the presence of at least one double bond.

Olefin copolymers. Polymers made by the interaction of two or more kinds of olefin monomers; butylene and propylene are examples.

Oleic liquid. A term used to indicate an organic liquid that usually connotes an immiscibility with water.

Oleophilic. Having an affinity for, attracting, adsorbing, or adsorbing oil.

Oligomer. Polymer made up of two, three, or four monomer units.

One-sided confidence limit. The upper limit on a parameter of a distribution.

One-sided tolerance limit. The upper limit on observations from a specified distribution.

Optical spectroscope. Optical instrument consisting of a slit, collimator lens, prism or grating, and a telescope or objective lens which produces an optical spectrum arising from emission or absorption of radiant energy by a substance for visual observation.

Optical spectroscopy. Production, measurement, and interpretation of optical spectra arising from either emission or absorption of radiant energy by a substance for visual observation.

Order. (1) The regular arrangement or pattern assumed by atoms or other chemical units in a space lattice; (2) classification of chemical reactions, in which the order is described as first, second, third, or higher, according to the number of molecules (one, two, three, or more) that enter into the reaction.

Order statistics. Sample values arranged in increasing order.

Organic acid. A chemical compound with one or more carboxyl radicals (COOH) in its structure.

Organic carbon partition coefficient (K_{oc}). The measure of the tendency for organic chemicals to be adsorbed by soil and sediment, expressed as the milligrams of chemical adsorbed per kilogram of organic carbon divided by the milligrams of chemical dissolved per liter of solution.

Organic chemicals. Chemical compounds of carbon, excluding carbon monoxide, carbon dioxide, carbolic acid, metallic carbides, metallic carbonates, and ammonium carbonate.

Organic chemistry. The study of the composition, reactions, and properties of carbon-chain or carbon-ring compounds or mixtures thereof.

Organic horizon. A soil horizon that contains more than 30% organic matter if the mineral fraction is more than 50% clay, or more than 20% organic matter if the mineral fraction has no clay.

Organic matter. A substance that contains carbon combined with hydrogen along with other elements.

Organic nitrogen. Nitrogen combined in organic molecules such as proteins, amines, and amino acids.

Organic polymers. Drilling fluid additives comprised of long-chained, heavy organic molecules which are used to increase drilling rates and drilling fluid yields, thereby decreasing operational costs.

Organic salt. The reaction product of an organic acid and an inorganic base.

Organic soils. A generic term applied to a soil that consists of organic matter, such as peat soils, muck soils, and peaty soil layers.

Organic vapor analyzer. A field monitoring device used to determine the concentrations of organic compounds in air using flame ionization detection (FID) or photoionization detection (PID) systems.

Organochlorines. Organic compounds containing tin which are formulated to act as anti-fouling agents in paints used on the hulls of boats and ships.

Organoleptic effects. Affecting one or more of the sensory organs. Usually refers to tests and odor effects in water.

Orifice plate. Flow measurement device for liquids or gases which uses a restrictive orifice plate consisting of a machined hole that produces a jet effect. Typically the orifice meter consists of a thin plate with a square-edged, concentric, and circular orifice. The pressure drop of the jet effect across the orifice is proportional to the flow rate. The pressure drop can be measured with a manometer or differential pressure gauge.

Orthophosphate. An acid or salt containing phosphorus as PO_4.

Oscillation. Periodic movement to and from or up and down.

Oscillator. Instrument used to produce an alternating current by converting a direct current source to a periodically varying electrical output.

Oscillograph. An instrument that graphically records the oscillations or changes in an electric current.

Osmosis. The process of diffusion of a solvent through a semi-permeable membrane from a solution of lower concentration to one of higher concentration.

Osmotic potential. The amount of work that must be done per unit quantity of pure water in order to transport reversibly and isothermally an infinitesimal quantity of water from a pool of pure water, at a specified elevation and atmospheric pressure, to a pool of water identical in composition to the equilibrium soil solution (at the point under consideration), but in all other respects being identical to the reference pool.

Osmotic pressure. The hydrostatic pressure at equilibrium, when solvent molecules are crossing the membranes in both directions at equal rates.

Osmotic water transport. The transfer of water from product compartments through a membrane into concentrating compartments.

Outlet. The point that a lake or pond discharges into a stream that drains it.

Outlier. An observation that does not conform to the pattern established by other observations in the data set.

Outwash. (1) Stratified accumulation of water-deposited glacial drift laid down by the meltwater streams from a melting glacier; (2) water-deposited material carried and laid down by streams.

Outwash plain. A plain formed by material deposited by water from a glacier flowing over a more or less flat surface of a large area.

Outwash sand. Stratified sediment (usually sand and gravel) removed from a glacier by meltwater streams and deposited beyond the active margin of a glacier.

Overland flow. The flow of snowmelt or rainwater over the land surface toward stream channels.

Oxalic acid ($HOOCCOOH(2H_2O)$). A poisonous, transparent crystal which is soluble in water, alcohol, and ether. It is used as a chemical intermediate, a bleach in polishes, and a rust remover.

Oxamyl ($C_7H_{13}N_3O_3S$). A white, crystalline compound used to control pests of tobacco, ornamentals, fruits, and other crops.

Oxic horizon. Subsurface soil that contains at least 15% clay and is 12 inches or more in thickness.

Oxidation. (1) Addition of oxygen to a compound; (2) loss of electrons by a constituent of a chemical reaction; (3) change in a compound caused by an increase in the proportion of the electronegative part or the change of an element or ion from a lower to a higher positive valence.

Oxidation number. The numerical charge on the ions of an element.

Oxidation-reduction potential. The potential required to transfer electrons from the oxidant to the reductant.

Oxidation-reduction reaction. An oxidizing chemical change, where the element's positive valence is increased (electron loss), accompanied by a simultaneous reduction of an associated element (electron gain).

Oxidation state (oxidation number). The number of electrons to be added (or subtracted) from an atom in a combined state to convert it to elemental form; the degree to which an element is oxidized.

Oxide box. A device used to remove sulfur impurities from manufactured gas which contained hydrated iron-oxide (usually dispersed on a bed of wood shavings, sawdust, peat, or similar material) which reacted with hydrogen sulfide and other trace contaminants in the gas.

Oxide cutan. A thin layer of metal oxide, usually iron or manganese, on the face of the soil particle.

Oxidizing acids. An acid (e.g., HNO_3) which tends to lose electrons in a reaction.

Oxidizing agent. A substance that can add electrons.

Oxygen demand. The quantity of oxygen utilized in the biochemical oxidation of organic matter in a specified time, at a specified temperature, and under specified conditions.

Ozonation. A chemical treatment process where the oxidation of the compound is achieved with ozone as the oxidizing agent.

Ozone. Oxygen in molecular form that consists of three oxygen atoms (O_3). Ozone is the most reactive form of oxygen and is a powerful oxidizing agent, much more active than ordinary oxygen, which is used for bleaching oils, waxes, ivory, flour, paper pulp, and starch and for disinfecting drinking water.

P

Packer. In well drilling, a device lowered in the lining tubes which swells automatically or can be expanded by manipulation from the surface at the correct time to produce a watertight joint against the sides of the bore hole or the casing to exclude water from higher horizons.

Packer test. An aquifer test in which two inflatable seals (packers) are placed in an open borehole to prevent the movement of water in the well while the hydraulic conductivity of the adjacent formation is measured.

Pahoehoe. Solidified lava that is characterized by a smooth, billowy, or ropy surface and having a skin of glass a fraction of an inch to several inches thick.

Palladium oxide (PdO). An amber or black-green powder that is soluble in dilute acids and is used in chemical synthesis as a reduction catalyst; also known as palladium monoxide.

Palmitoleic acid ($C_{16}H_{30}O_2$). An unsaturated fatty acid which is used as a standard in chromatography; also known as *cis*-9-hexadecenoic acid.

Paper chromatography. A procedure for the analysis of complex chemical mixtures.

Paraconformity. A time break in a sequence of depositions where the unconformity is parallel to the strata above and below it.

Paraffin. A major group of aliphatic hydrocarbons derived from petroleum; they may be either straight or branched chain.

Paraformaldehyde ($(HCHO)_n$). A polymer of formaldehyde which is used as a disinfectant, fumigant, and fungicide and to make resins.

Paraldehyde ($C_6H_{12}O_3$). An acetaldehyde polymer which is miscible with most organic solvents and soluble in water. It is used as a chemical intermediate, in medicine, and as a solvent.

Paraldol ($CH_3CHOHCH_2CHO$). A water-soluble crystal used as a chemical intermediate, in manufacturing resin, and in cadmium plating baths.

Parameter. In statistics, an unknown constant associated with a population.

Paraquat ($[CH_3(C_5H_4N)_2CH_3](2CH_3So_4)$). A yellow, water-soluble solid which is used as an herbicide.

Parent compound. A chemical compound which is the basis for one or more derivatives; for example, ethane is the parent compound for ethyl alcohol and ethyl acetate.

Parent name. That portion of the name of a chemical compound from which the name of a derivative comes.

Parinol ($C_{18}H_{12}C_{12}NO$). A white to pale yellow solid used as a fungicide to control powdery mildew on flowers, nonbearing apple trees, and grape vines.

Part B permit. The second, narrative section submitted by generators in the RCRA permitting process; it covers detailed procedures followed at a facility to protect human health and the environment.

Partial correlation analysis. A technique that estimates a measure of the relationship between two variables while controlling for the effects of other variables. These effects are controlled by removing the linear relationship with other variables before calculating the correlation coefficients between the two variables of interest.

Partial correlation coefficient. A measurement of the relationship between two variables while controlling possible effects of other variables.

Partial molar volume. The volume of a solution or mixture related to the molar content of one of the components within the solution or mixture.

Partial percent retained. In sieve analysis of a soil, it is the weight in grams retained on a sieve divided by the weight in grams of a sample used for a given series of sieves.

Partial pressure. The portion of total vapor pressure in a system due to one or more constituents in the vapor mixture.

Partial vacuum. The description of a space condition in which the pressure is less than atmospheric.

Particle. A term used to refer to individual aggregates of matter having characteristic properties and structure.

Particle size analysis (mechanical analysis). A laboratory technique that measures the percentages of sand, silt, and clay in a soil.

Particle tracking. A mathematical approach used to trace the movement of imaginary particles in a flow field. This technique is commonly used in contaminant transport modeling.

Particulates. Solid particles suspended in air, such as soot, ash, and dust.

Partition. Term describing a solvent system in which the solute is divided between two components of the liquid phase. The ratio of their concentrations (partition coefficient) is constant for a given substance.

Partition coefficient (K). In the equilibrium distribution of a solute between two liquid phases, it is the constant ratio of the solute's concentration in the upper phase to its concentration in the lower phase.

Parts per million (ppm). A unit measure of concentration that is a weight-to-weight ratio. For example, a 1 ppm concentration of arsenic represents 1 gram of arsenic to 999,999 grams of other material.

Passive remediation. Natural degradation of chemicals through physical, chemical, and biological processes.

Pathline. A general term that refers to a flow path. It describes the route that a discrete particle of water follows through a region of flow during a steady or transient event.

pE. The measurement of the oxidation-reduction state of a solution at equilibrium; the pE of a solution is a measurement of the negative logarithm of the electron activity of the solution.

Peak width. In a gas chromatogram, the width of the base of a symmetrical peak.

Pearl hardening. A commercial name for a crystallized grade of calcium sulfate.

Peat. An acidic, dark-colored, and coarsely fibrous unconsolidated soil consisting of 96 to 99% decomposed plant matter.

Peat bog. A soft, wet, spongy area that consists chiefly of decayed or decaying moss and other vegetable matter.

Peat soil. Organic soil formed by the accumulation in wet areas of the partially decomposed remains of vegetation and having less than 20% mineral material.

Pebble. A small stone, worn smooth by the action of water, ice, or sand, which is between 4 and 64 millimeters in diameter.

Pedon. The smallest volume that can be recognized as a soil.

Pelletizing. In ore preparation, it is the process of agglomerating fine magnetic concentrations of taconite ores. The ore is ground, sized, and mixed with water and binder, then rolled into small balls.

Pellicular water. Water adhering as films to the surfaces of grains of water bearing material after gravitational-dependent water has been drained. It occurs as wedge-shaped bodies at the junctures of soil grains in the unsaturated zone.

Peltier coefficient. The ratio of the theoretical rate at which heat is evolved or absorbed at a junction of two dissimilar metals (Peltier effect) to the current passing through the junction.

Peltier effect. Heat which is evolved or absorbed after allowing for resistance at the junction of two dissimilar metals (as in a thermocouple) carrying a small current and which is dependent upon the direction of the current.

Pennsylvanian. A geologic epoch between 320 and 286 million years ago.

Pentabasic. A molecule that has five hydrogen atoms that may be replaced by metals or bases.

Pentachlorethane ($CHCl_2CCl_3$). A colorless, water-insoluble liquid which is used as a solvent to degrease metals; also known as pentalin.

Pentachlorophenol (C_6Cl_5OH). A toxic white powder which is soluble in alcohol, acetone, ether, and benzene and is used as a fungicide, bactericide, algicide, herbicide, and chemical intermediate.

Pentadecane ($C_{15}H_{32}$). A colorless, water-insoluble liquid which is used as a chemical intermediate.

Pentadentate ligand. A chelating agent that has five groups capable of attachment to a metal ion.

Pentaerythritol (($CH_2oH)_4C$). A white crystalline solid used to make the explosive pentaerythritol tetranitrate (PETN) and to manufacture alkyl resins and other coating compounds.

Pentaerythritol tetrastearate ($C(CH_2OOCC_{17}H_{35})_4$). A wax used in polishes and textile finishes.

Pentafluoride. A chemical compound onto which five fluoride atoms are bonded.

Pentane ($CH_3(CH_2)_3CH_3$). A colorless, flammable, water-insoluble hydrocarbon liquid which is soluble in hydrocarbons and ethers and is used as a chemical intermediate, solvent, and anesthetic.

Pentanediol ($HOCH_2(CH_2)_3CH_2OH$). A colorless liquid used as a hydraulic fluid, lube-oil additive, and antifreeze and in the manufacture of polyester and polyurethane resins.

Peptization. The dispersion of a fluid caused by the addition of electrolytes or other chemicals.

Percent recovery. The percentage of a spiked analyte that is recovered. For spiked recoveries, the relationship is expressed as:

$$\frac{(\text{average of spike anaysis results} - \text{sample concentration})}{\text{spike amount}}$$

This ratio is then multiplied by 100.

Percentage base saturation (PBS). The cation exchange capacity of a soil that is calculated by summing the exchangeable bases and exchange acidity. It is expressed as:

$$PBS = \frac{\sum \text{exchangeable bases} \times 100}{\sum \text{exchangeable bases} + \text{exchange acidity}}$$

Percentile. A value below which a certain percentage of observations fall; for example, the 30th percentile is the value below which 30% of the observations fall.

Perched water. Unconfined groundwater separated from an underlying main body of groundwater by an unsaturated zone.

Perched water table. An unconfined water surface that is separated from an underlying body of groundwater by an unsaturated zone.

Percolate. The movement of water seeping or filtering through the soil without a definite channel.

Percolation. Downward movement of water through soil, especially the downward flow to water in saturated or nearly saturated soil at hydraulic gradient of the order of 1.0 or less.

Percolation test. An *in situ* test measuring the suitability of a soil for a sewage disposal system (leachfield).

Performance evaluation samples (PE). A sample with known analytes at verifiable concentrations which is given to a laboratory for evaluation of the ability of the analytical laboratory to generate accurate and defensible data.

Period. (1) Time interval required to complete a cycle; (2) smallest increment of the independent variable for which a function repeats itself.

Peristaltic pump. Low-volume suction pump in which the compression of a flexible tube by a rotor results in suction development.

Permeability (k). The property or capacity of a porous rock, sediment, or soil for transmitting a fluid; it is a measure of the relative ease of fluid flow under unequal pressure.

Permeability test. A test used to determine the hydraulic conductivity of the aquifer formation near a well screen. It is generally conducted by displacing the water level in a well and monitoring the rate of recovery of the water level as it returns to equilibrium. Various methods of analysis are available to calculate the hydraulic conductivity from these data.

Permeation rate. The rate at which a hazardous chemical moves through a protective material; it is measured in milligrams per square meter per second ($mg/m^2/sec$).

Permissible exposure limits (PELs). The concentration established by the Occupational Safety and Health Administration (OSHA) which represents the acceptable concentration of airborne chemicals in the workplace. PELs differ from threshold limit values (TLVs) in that they represent enforceable workplace air concentrations, specific to each chemical, to which workers may be exposed for 8 hours a day, 5 days a week, for a period of 47 years, without adverse health effects.

Pesticide. Substance or mixture of substances intended for preventing, destroying, repelling, or mitigating any pest. Any substance or mixture of substances intended for used as a plant regulator, defoliant, or desiccant.

Petrochemical. An intermediate chemical derived from hydrocarbon liquids, natural gas, or petroleum.

Petroleum. A complex mixture of hydrocarbon compounds with minor amounts of nitrogen, oxygen, and sulfur as impurities.

Petroleum gas. Any of a number of flammable gases which are obtained by refining of petroleum and are used for organic synthesis and in liquefied form as fuels.

Petroleum spirits (mineral spirits, white spirits, or Stoddard solvents). A hydrocarbon derived from the light distillate fractions during the crude-oil refining process. They are composed of the C_6 to C_{11} compounds, with the majority of the relative mass composed of C_9 to C_{11}. Petroleum spirits are composed of the following general classes of compounds: 50% paraffins, 40% cycloparaffins, and 10% aromatics.

pF. (Obsolete); a common logarithm of the height of a water column in centimeters which is equivalent to the soil moisture tension. pF is often expressed as suction (negative pressure) or on a potential energy (per unit mass) basis.

pH. A measure of the acidity or alkalinity of a solution, numerically equal to 7 for neutral solutions, which expresses the hydrogen-ion activity of a solution. pH is theoretically represented as:

$$pH = -\log[H^+], \text{ or}$$

$$pH = \log 1/[H^+]$$

Phase. Any portion of a physical system separated by a definite physical boundary from the rest of the system.

Phenol. A class of highly soluble aromatic organic compounds based on the substitution product of phenol.

Phenolic. General name for a class or group of chemicals derived from benzene by substitution of one or more hydroxyl groups, and in some cases methyl groups, as well.

Phenolic compounds. Hydroxy derivatives of benzene.

Photochemistry. A subdivision of chemistry devoted to the chemical changes induced by various wavelengths of radiation, often brought about through the agency of molecular fragments known as free radicals.

Photodegradation. The phenomenon whereby ultraviolet radiation attacks a chemical bond or link in a polymer or chemical structure.

Photogrammetry. (1) Science dealing with obtaining reliable measurements from photographic images; (2) process of making maps and scale drawings from aerial photographs.

Photoionization detector. A detector that uses an ultraviolet light source to ionize individual constituents. Gaseous contaminants are ionized as they emerge from the column, and the ions are then attracted to an oppositely charged electrode, resulting in an electrical current which is amplified and measured on a numerical scale (meter readout) or recorded on paper (strip chart).

Photolineament. Any line on an aerial photograph that is structurally controlled, including alignment of separate photographic images such as stream beds, trees, or bushes. The term is applied to lines representing beds, lithologic horizons, mineral bandings, faults, joints, unconformities, and rock boundaries.

Photolithography. The process using a photosensitive emulsion and light to transfer a pattern or image from a mask to a wafer.

Photolysis. Decomposition or cleavage of a chemical compound induced by radiation of certain wavelengths.

Photometer. An instrument that measures the light intensity or the degree of light absorption.

Photosynthesis. The process by which plants convert solar radiation into chemical energy.

Phototrophs. Organisms that use light to generate energy (by photosynthesis) for cellular activity, growth, and reproduction.

Phreatic surface. The upper surface of the saturation zone, except where that surface is formed by an impermeable body.

Phreatic zone. Synonymous with the term *saturated zone*.

Physical chemistry. That portion of science that deals with laws or generalizations related to chemical phenomena.

Phytotoxic. Poisonous to plants.

Pickling. Removal of scale from steel and iron by immersion in hot hydrochloric or sulfuric acid.

Picocurie (pCi). A unit of radioactivity equal to 10^{-12} curie.

Piezometer. (1) Any of several instruments used to measure the liquid pressure in soil or other porous material; (2) device to measure the rise of groundwater to the piezometric surface.

Piezometer nest. Multiple well completions in the same borehole with each well screened over a different interval.

Piezometric surface. An imaginary surface that everywhere coincides with the static level of the water in an aquifer.

Piezometric surface map. A contour map of the imaginary surface to which the water in an artesian aquifer will rise.

Pilot test. Operation of a small-scale version of a larger system to gain information relating to the anticipated performance of the larger system. Pilot test results are typically used to design and optimize the larger system.

Piping. Erosion by percolating waters or seepage in a layer of subsoil resulting in caving and the formation of tunnels or pipes through which the soluble or granular material is removed.

Piston pump. A pump that consists of a piston rod, cylinder, and check valve. Piston pumps force water to the surface through positive displacement.

Pitch (rake). Angle measured in a specified plane, between a line and the horizontal.

Pitot tube. An instrument used to measure the relative speed of a fluid.

Pitting factor. The depth of the deepest pit resulting from corrosion divided by the average penetration as calculated from weight loss.

Planar area. The area of a region in the x,y-coordinate plane.

Planimetric map (line map). Map representing only the horizontal position of features.

Plank's constant (h). A physical constant equal to 6.63×10^{34} joules per degree K.

Plank's Law. A law that provides the intensity distribution of energy emitted by a blackbody as a function of wavelength and temperature. It is mathematically expressed as:

$$E\gamma = C_1/\gamma^5[\text{exponent } (C_2/\gamma T) - 1]$$

where

  $E\gamma$ = energy flux emitted by a particular wavelength range.
  C_1 and C_2 = constants.
  T   = temperature.

Planosol. Soils with exuviated surface horizons underlain by claypans or fragipans. These soils are developed on nearly flat or gently sloping uplands in humid or subhumid climates.

Plastic. A polymer capable of plastic flow or deformation during a stage in its manufacture which allows it to be molded into a desired shape. In soil morphology, it is a qualitative term that describes the plasticity of a soil; it denotes a soil that, when rolled between the hands, forms a long wire (>1 centimeter) of soil and requires moderate pressure to deform the molded soil.

Plastic limit (PL). The lower limit of the plastic state of a soil. It is that water content at which the soil begins to crumble when rolled into thin threads.

Plastic soil. A soil that will deform without shearing (typically silts or clays). Plasticity characteristics are measured using a set of parameters known as Atterberg Limits.

Plasticity index (PI). The range in water content between the liquid limit and the plastic limit of a soil.

Plasticizer. (1) Substance incorporated into a plastic or elastomer for the purpose of improving the material's flexibility and ability to be processed; (2) nonvolatile organic liquid of medium viscosity which is added to rubber and plastic mixtures to act as an internal lubricant and softener.

Platy soil structure. Soil aggregates with thin vertical axes and long horizontal axes.

Playa (dry lake). A flat or nearly flat portion of an enclosed basin or temporary lake without an outlet.

Pleistocene. A geologic epoch that occurred between 1.6 million and 11,000 years ago.

Plume, groundwater. A mass of contaminated water extending outward from the source; the zone of contamination that contains compounds in the dissolved phase.

Plunge. The vertical angle between a line and horizontal.

Plutonic. Of igneous intrusive origin; usually applied to similar deposits containing other metals such as tin, platinum, or tungsten.

Podzolization. Chemical migration of aluminum, iron, and/or organic matter from a soil horizon.

Point source. A single, concentrated identifiable source of pollutants, such as a sewer outfall or factory smokestack.

Poisson's ratio. An elastic constant (does not exceed 0.5) which is the ratio of the lateral unit strain to the longitudinal unit strain in a body that has been stressed longitudinally within its elastic limit.

Polar compound. A molecule which has, or can acquire, electrical charges which enable it to conduct electricity.

Polarization. The state of an electromagnetic wave in which the electric and magnetic field vibrates in a straight line in the plane perpendicular to the direction of wave propagation.

Pollyallomer. A special type of copolymer of two olefinic compounds, such as propylene and ethylene, in a ratio of about 20 to 1.

Polyaromatic hydrocarbon (PAH). Aromatic hydrocarbons containing more than one fused benzene ring.

Polychlorinated biphenyl (PCB). A chemical substance limited to the biphenyl molecule that has been chlorinated to varying degrees or any combination of substances that contain 50 ppm (on a dry-weight basis) or greater of such substances.

Polyelectrolyte. Ionically active, electrically conducting high polymer, either natural or synthetic, used for water conditioning, colloidal clay dispersion, and inhibition of calcium carbonate deposition.

Polyethylene. A thermoplastic polymer composed of a synthetic crystalline polymer of ethylene; polyethylene may be of low density (branched) or high density (linear).

Polygonal soil. A more or less regular-sided ground surface pattern formed by frost action, thawing, or ground ice wedges. It is common in permafrost soils and is indicative of poor soil drainage.

Polymer. A substance formed by the union of two or more molecules of the same kind linked end to end into another compound having the same elements in the same proportion but a higher molecular weight and different physical properties.

Polymerization. A chemical reaction in which simple materials, either one or more, are converted to a complex material which possesses properties entirely different from the original materials used at the start of the reaction.

Polynomial smoothing. A procedure that uses an nth-order polynomial fitted to a time series to estimate a polynomial trend in the data set.

Polynuclear aromatic hydrocarbon (PNA). Synonymous with polyaromatic hydrocarbon.

Polyvinyl chloride (PVC). A thermoplastic polymer produced by polymerizing vinyl chloride manometer or vinyl chloride/vinyl acetate monomers; PVC is usually rigid and contains plasticizers.

Pore. As applied to stone, soil, or other material, any small interstice or open space, generally one that allows the passage or adsorption of liquid or gas.

Pore space. Open space in rock or granular material.

Pore volume. Volume of water (or air) that will completely fill all of the void space in a given volume of porous matrix. Pore volume is equivalent to the total porosity. The rate of decrease in the concentration of contaminants in a given volume of contaminated porous media is directly proportional to the number of pore volumes that can be exchanged (circulated) through the same given volume of porous media.

Pore water (interstitial water). Water occupying an open space between solid soil particles.

Porosity (n). The percentage of the bulk volume of a rock or soil that is occupied by interstices, whether isolated or connected. It is expressed as:

$$n = (V_v/V) \times 100\%$$

where

n = porosity.
V_v = volume of void space.
V = total volume of material (solid and void space).

Representative porosity values for soils are

Unconsolidated Sediments	Porosity (%)
Glacial till	10–20
Sand and gravel	20–35
Gravel	25–50
Coarse sand	33
Medium sand	35–40
Fine sand	45–52
Silt	35–50
Clay	33–60

Porosity, effective (n_e). The amount of interconnected pore space available for fluid transmission which is expressed as a percentage of the total volume occupied by the interconnecting interstices.

Potential energy. The energy possessed by a body of matter due to its position or condition.

Potentially active fault. Any fault that shows evidence of surface displacement during Quaternary time (last 1.6 million years).

Potentiometer. Instrument used for the measurement of an electrical potential by comparison with a known potential difference.

Potentiometric surface. Replaces the term *piezometric surface*; a surface that represents the static head. As related to an aquifer, it is defined by the levels to which water will rise in wells. Where the head varies appreciably with depth in the aquifer, a potentiometric surface is meaningful only if it describes the static head along a particular specified surface or stratum in that aquifer. The water table is a particular potentiometric surface.

Power. In statistics, the power of a test is a measure of the ability of the test to detect a difference of specified size from the null hypothesis.

Pozzolanic process. Method of solidification/stabilization in which waste is mixed with fine-grained siliceous materials such as fly ash or cement kiln dust to produce a solid.

ppm. Parts per million, 1×10^{-6}. A convenient means of expressing very low concentrations of a substance in a mixture or a low-level contaminant in a pure product.

Precambrian. A geologic era that ended about 570 million years ago.

Precipitator. A device used in factory stacks which utilizes an electrostatic field to introduce an electric charge onto solid or liquid suspended particles in a gas stream so that the particles migrate to one electrode from which they can be collected.

Precision. The degree to which a measurement is reproducible.

Press filter. A press operated mechanically for partially dewatering sludge.

Pressure. The total load or force acting on a surface, usually expressed as a unit of force per unit area.

Pressure gauge. A device for registering the pressure of solids, liquids, or gases.

Pressure gradient. A pressure differential in a given medium (e.g., water or air) which tends to induce movement from areas of higher pressure to areas of lower pressure.

Probability sampling. The use of specific methods or random selection of population units for measurement.

Procaryotes. A cellular organism in which the nucleus has no limiting membrane.

Producer gas. A gas rich in carbon monoxide, hydrogen, and nitrogen (heating value of about 150 Btu per cubic foot) made by reacting coal or coke with steam and air.

Protective anode. A metallic object designed to corrode in place of the object it is designed to protect.

Protozoa. Single-celled, eucaryotic microorganisms without cell walls. Most protozoa are free living, although many are parasitic. The majority of protozoa are aerobic or facultatively anaerobic heterotrophs.

Psi. Pounds per square inch absolute; the absolute thermodynamic pressure, measured by the number of pounds force exerted upon an area of one square inch.

Psig. Pounds per square inch gauge; 0 psig = 14.969 psi (absolute) = 1.0 atmosphere.

Psychrometer. (1) Instrument used to determine the relative humidity and vapor tension of the atmosphere. (2) In soil physics, an instrument used to measure soil humidity by means of a thermocouple which is cooled below the dew point. The water on the thermocouple evaporates, causing the junction temperature to be depressed below the ambient temperature. The wet bulb temperature depression persists until all the water has evaporated; the thermocouple then returns to the ambient temperature.

Psychrometry. Measurement of the humidity of a system.

Pugmill. A chamber in which water and soil are mixed together. Typically mixing is aided by an internal mechanical stirring/kneading device.

Pulping. A term used primarily in the paper industry to refer to any of a number of methods of preparing woody fibers for the papermaking operation. The chief purposes of pulping are the separation and removal of lignin from cellulosic components of the wood and the softening of the fibers. These are accomplished in the following ways: (1) chemical pulping, which involves digestion or cooking of softwood pulp in either an acidic (sulfite pulp) or a basic medium (sulfate or Kraft pulp); (2) semi-chemical and chemi-mechanical pulping, designed chiefly for hardwoods in which chemical and physical methods are combined; and (3) mechanical or groundwood pulping, mainly for newsprint, in which the pulp is formed by abrading the wood with a rotating stone.

Pulse height analyzer. An instrument used to measure the pulse height whenever the amplitude of the pulse is proportional to the energy dissipation in a detector.

Pump. Mechanical device used to cause flow for raising or lifting water or other fluid or for applying pressure to fluids.

Pump test. A test made by pumping a well for a period of time and observing the change in hydraulic head in adjacent wells. A pump test may be used to determine the degree of hydraulic interconnection between different water-bearing units, as well as the recharge rate of a well.

Pumping test. A test that is conducted to determine aquifer or well characteristics.

Purge (wells). The process of pumping out of well water to remove drilling debris or impurities; purging is also conducted to bring fresh groundwater into the casing for sample collection. The latter is a means of collecting a representative water sample from the aquifer being investigated.

Purgeable organic. An organic chemical with a high vapor pressure that can be removed from water by bubbling a nonreactive gas such as helium in the water.

Purged water. Water from wells undergoing evacuation or being used for aquifer testing.

Purification. Removal of objectionable matter from water by natural or artificial methods.

Purifier box. A common term for oxide boxes associated with manufactured gas plants.

PVC. *See* Polyvinyl chloride.

Pycnometer. A container used to determine the density of a liquid or soil having a specific volume; a thermometer is often included to indicate the temperature of the contained substance.

Pyrheliometer. An instrument for measuring the intensity of incoming radiation from the sun or the sky.

Pyrite. Iron sulfide.

Pyrolysis. The thermal decomposition of organic material in an oxygen-deficient atmosphere.

Q

Quadrant search method. A method used for gridding data in which the original data points used to estimate a grid element are found by dividing the area around a grid element into four quadrants. This search method finds the nearest points in each quadrant around the element estimate.

Quantitative analysis. The analysis of a gas, liquid, or solid sample or mixture for the purpose of determining the precise percentage composition of the sample in terms of elements or compounds.

R

Radical. In organic chemistry, a group of atoms that occurs repeatedly in a number of different compounds. Among the more common radicals are the hydroxyl $(OH)^-$, nitrate $(NO_3)^+$, ammonium $(NH_4)^+$, and sulfate $(SO_4)^=$ compounds.

Radio frequency (RF) heating. Remediation process in which soil is heated by exciting an array of electrodes with an RF generator to volatilize and remove volatile organic compounds.

Radioisotopes. Radioactive isotopes of an atom, usually artificially produced.

Radionuclide. Radioactive nuclide; an atom having an unstable nucleus that spontaneously disintegrates and emits alpha, beta, or gamma radiation.

Radius of influence. Radial distance from the center of a well bore to the point where there is no lowering of the water table or potentiometric surface (the edge of its cone of depression).

Raoult's Law. A law stating that the extent of the physical blocking effect or depression of the vapor pressure is directly proportional to the concentration of the particles in solution.

Random sampling error. Variation in an estimated quantity due to the random selection of units for measurement.

Reactivity. Susceptibility of a substance to undergoing a chemical reaction or change that may result in dangerous side effects, such as an explosion, burning, and corrosive or toxic emissions.

Reagent. A substance used in chemistry to detect, measure, or produce another substance.

Reagent blanks. Reagent blanks (or solvent blanks) are used when solvent extracts are analyzed, as in the case of PCBs or other semi-volatile compounds. A reagent or solvent blank is simply an injection of "clean" solvent to ensure that contaminants are not present as impurities in the solvent being used for extractions. Any contamination introduced via the syringe or any other transfer vehicle will also be detected by a solvent blank. As in the case of syringe blanks, the blank chromatogram should be free of peaks other than the solvent peak itself.

Reagent water. Water in which an interferant is not observed at the method detection limit of the analyte that is of interest.

Real density. In soil mechanics, the mass per unit volume of oven-dried solid soil matter, pore space excluded.

Recalcitrant. Synonymous with unreactive, nondegradable; refractory.

Recharge. Addition of water to the zone of saturation.

Recharge area. Replenishment of an aquifer by a natural process such as addition of water at the ground surface or by an artificial system such as addition through a well.

Rectifier. An electronic device that changes alternating current to direct current.

Redox (reduction/oxidation). A chemical reaction in which an atom or molecule loses electrons to another atom or molecule. Oxidation is the loss of electrons; reduction is the gain in electrons.

Reducing agent. A substance that can give up electrons. A chemical that combines with oxygen and removes it from a substance.

Reduction/oxidation. A chemical treatment process where the oxidized state of one reactant is raised while that of another is lowered. This process destroys or lessens the toxicity of many organics and heavy metals.

Refractory. A term used by materials engineers to designate any substance or mixture which withstands extremely high temperatures without losing its shape or chemical identify. These products are used in furnaces, jet engines, rocket propulsion systems, and similar high-temperature environments. Tungsten, tantalum, and so-called superalloys are examples of refractory metals; silicon carbide, clay, alumina, magnesium oxide, borosilicate glasses, and porcelain are refractory ceramics.

Refractory index. A measure of the ability of a substance to be biodegraded by bacterial activity; the lower the refractory index, the greater the biodegradability.

Relative compaction. The amount of compaction relative to the moisture density curve or compaction curve.

Relative coordinate. A coordinate that is referenced to a temporary origin or position.

Relative density. The ratio of the difference between the void ratio of a cohesionless soil in the loosest state and any given void ratio to the difference between its void ratios in the most loose and dense states.

Relative error. The absolute error divided by the true value of a parameter. Because the true value of a parameter is usually unknown and the absolute error is an estimate, the relative error is also an estimate.

Relative humidity. The ratio of the quantity of water vapor present in air to the quantity that would saturate it at any specific temperature.

Representative sample. A sample assumed not to be significantly different from the population of samples available. In fuel leak investigation, samples are often selected to be representative of the worst-case situation.

Residual clay. Clay that has been formed in place by the weathering of rock.

Residual DNAPL (dense non-aqueous phase liquid). The DNAPL held in soil pore space by capillary forces.

Residual drawdown. The difference between the original static water level and the depth to water at a given distance during the recovery period of a well.

Residual fuel oil. The residue remaining from the distillation of petroleum that does not economically justify further processing.

Residual material. Unconsolidated and partly weathered parent material from soils presumed to have developed from the same kind of rock as that on which it lies.

Residual saturation. The fraction of total soil pore space filled with a liquid due to capillary forces; residual saturation in unsaturated soils typically ranges from 5 to 20% of the total pore volume.

Retaining wall. In engineering geology, a structure used to stabilize a slope.

Retardation. The preferential retention of contaminant movement in the subsurface due to sorption and solubility differences.

Retardation equation. A relationship that describes the retardation of a contaminant mass moving from a point source while undergoing adsorption. It is mathematically described by the following equation:

$$\varphi_a = 1/([1 + \rho_b/\eta]K_d)$$

where

φ_a = relative velocity of the groundwater.
ρ_b = bulk density.
η = porosity.
K_d = distribution coefficient.

Retardation factor. The variables in the retardation equation described by $1/([1 + \rho_b/\eta]K_d)$.

Retarder. A benzenoid compound, such as salicylic or benzoic acid, added to rubber and plastic mixes during processing; it is added in low concentrations to prevent cross-linking action (curing) from taking place at the temperatures prevailing during extrusion and other manufacturing steps.

Retention time. In gas chromatography, it is the time required for a compound to emerge from the column. The area under the peak for that compound on the chromatogram is proportional to the concentration of the compound in the sample.

Reticle. A piece of glass with a chrome pattern for several die, used in the photolithography process.

Retort gas. A gas rich in hydrogen and methane (heating value of about 500 Btu per cubic foot) made by carbonizing coal in a refractory-lined reactor, or retort.

Reverse osmosis (RO). A membrane separation technique which allows a solvent to be removed from a solution containing solutes by the application of a pressure-driven membrane process.

Revolving screen. A separation device consisting of a perforated drum which allows materials of various sizes to be graded and removed.

Reynolds number (R_e). A dimensionless number that expresses the ratio of inertial to viscous forces during fluid flow. It is commonly used to distinguish the boundary between laminar and turbulent flow. For flow through porous media, the R_e is described by:

$$R_e = \rho\varphi d/\mu$$

where

ρ = fluid density.
φ = fluid velocity.
d = a representative length for the porous media through which the fluid is flowing.
μ = fluid viscosity.

Darcy's Law is usually considered valid as long as R_e does not exceed a value between 1 and 10 (i.e., laminar flow).

Rigidity. The ratio of the shearing stress to the amount of angular rotation produced in a rock sample.

Riser pipe. A section of perforated well construction material used to connect the well screen with the ground surface. It is frequently made of the same material and has the same diameter as the well screen.

Risk assessment. The characterization of the potential adverse effects on human life or health or on the environment. A risk assessment can include a description of the potential adverse health effects based on an evaluation of results of epidemiological, clinical, toxicological, and environmental research. Extrapolation from those results is used to predict and estimate the type and extent of health effects in humans under given conditions of exposure.

Rock quality designation (RQD). A measure of the intactness of a rock core. It is the total combined length of all the pieces of intact core that are longer than twice the diameter of the core recovered during the core run divided by the total length of the core run. Typically, a core with a diameter of 2 inches is used. RQD is expressed as a percentage, and descriptive words may be used to supplement the range of RQD values.

Rotary drilling. A method of drilling wells in which the drill bit is rotated in the hole.

Roughness coefficient. A factor representing the effect of roughness of the confining material on the energy losses in the flowing water.

Running ground. Soil that flows into a tunnel from the floor, roof, or walls driven by water seepage. Typically the flow consists of cohesionless soil below the water table.

Runoff. The overland movement of water.

Rupture surface. The deepest surface along which movement occurs.

Rust. The reddish corrosion product formed by the electrochemical interaction between iron and atmospheric oxygen — ferric oxide Fe_2O_3. The reaction occurs most rapidly in moist air.

RQD (%)	Description
0–25	Very poor
25–30	Poor
30–75	Fair
75–90	Good
90–100	Excellent

S

Sacrificial. A term used by electrochemists to denote a type of galvanic corrosion in which a coating, most commonly zinc, is attacked in preference to the underlying metal. The coating is "sacrificed" in order to preserve the basis metal.

Salic horizon. A subsurface soil that is at least 15 centimeters thick and is salt enriched (i.e., 2 to 3%).

Saline water. Water containing dissolved salts, usually from 10,000 to 33,000 mg/L.

Salinity. Total amount of dissolved material in parts per million of total dissolved solids.

Salinization. An accumulation of soluble salts such as sulfates and chlorides of calcium, magnesium, sodium, and potassium in a soil horizon.

Salt. One of the products resulting from a reaction between an acid and a base.

Salt water intrusion. Invasion of a freshwater body by a saltwater body.

Saltation. Movement of soil and mineral particles by intermittent leaps from the ground when the particles are being moved by water or wind.

Sample standard deviation. The square root of the sample variance.

Sampling error. An error that arises from a sampling plan that includes only selected sampling units rather than the entire population.

Sand. Sediment particles having diameters between 0.062 and 2.0 millimeters.

Sandstone. A sedimentary rock containing more than 50% sand-sized particles that are predominately quartz. Examples are quartzose sandstones and orthoquartzites.

Sand boil. A cone-shaped deposit of sand formed during an earthquake when subsurface sand layers liquefy and then are blown to the surface through cracks.

Sandy clay. A soil of the U.S. Department of Agriculture textural class that contains 35% or more clay and 45% or more sand.

Sandy clay loam. A soil of the U.S. Department of Agriculture texture class that contains 20 to 35% clay, less than 28% silt, and 45% or more sand.

Sandy loam. A soil with a combination of sand, silt, and clay in the following percentages: sand (70–85%), silt (10–20%), and clay (10–20%).

Saponification. The reaction in which caustic material combines with fat or oil to produce soap.

Saturated hydrocarbons. Hydrocarbons in which adjacent carbon atoms are joined by a single covalent bond while all other bonds are linked by hydrogen.

Saturated rock. Rock that has all its interstices or void spaces filled with water.

Saturated soil. Soil that has all its interstices or void spaces filled with water, to the point at which runoff occurs.

Saturated zone. An underground geologic formation in which the pore spaces or interstitial spaces in the formation are filled with water under pressure equal to or greater than atmospheric pressure.

Scalar. A quantity that requires only a single value for specification; temperature and pressure are examples of scalars.

Scaling. The removal of oxides and/or corrosion products in the form of scales from a metallic or mineral surface.

Scanning curves. Curves between the wetting and drying curves in a soil moisture characteristic curve, which describes the wetting and drying values between different intermediate wetness values.

Scintillation count. The total number of light flashes produced in a phosphor by a given ionizing event.

Scintillation counter (Scintillation detector). A radiation detector which is used to measure ionizing radiation and consists of a scintillator combined with a photomultiplier.

Scintillator. An instrument used to convert radioactive energy into light. When an ionizing particle is absorbed in any one of several transparent scintillators, it is emitted as a pulse of visible or near-visible ultraviolet light.

Screen (well). The openings in a well casing that allows liquid to flow into the well.

Screened interval. That portion of a monitoring well that is open to the aquifer.

Scrubbers. Devices that remove pollutant gases and particulates from exhaust gases of power or manufacturing plants.

Secondary interstices. An interstice in rock that developed by processes that affected the rock after it was formed.

Secondary standard. In chemistry, a solution that has been standardized against a primary standard.

Second-order reaction. A reaction in which the reaction rate is proportional to the square of the concentration of one of the reactants or to the product of the concentration of two different reactants.

Sedentary soil (residual soil). Soil formed in place without removal from the site of the original rock from which it was formed.

Sediment. A deposit of solid material (or material in transportation which may be deposited) composed of any medium on the surface of the Earth.

Sedimentary rock. Rock which originated as a sediment. The sediment may have been transported by wind, water, or ice and carried in the form of solid particles (sand, gravel, clay) or in solution (rock, salt, gypsum, some calcareous sediments). Sedimentary rocks (unless still unconsolidated) have been indurated by cementation or by recrystallization.

Seebeck effect. Phenomenon occurring when two dissimilar metals are welded at one end; heating of this welded juncture results in a voltage on the free ends.

Seepage. The movement of water or other fluid through a porous material such as soil.

Seepage velocity (v_s). The rate at which a liquid moves through a porous media. For Darcian flux, it is equal to the saturated hydraulic conductivity times the hydraulic gradient divided by effective porosity.

Selective membrane. Material that is preferentially selective to passage of either cations or anions in solution.

Semiconductor A medium which has electrical properties between those of a conductor, such as copper, and those of an insulator, such as rubber.

Semiperched water. Groundwater that has a greater pressure head than an underlying body of groundwater from which it is not completely separated hydraulically.

Sensitivity. The ability of a method to distinguish significantly between small differences in concentration of analyte.

Sentinel well. A groundwater monitoring well situated between a sensitive receptor downgradient and the source of a contaminant plume upgradient. Contamination should be first detected in the sentinel well which serves as a warning that contamination may be moving closer to the receptor. The sentinel well should be located far enough upgradient of the receptor to allow enough time before the contamination arrives at the receptor to initiate other measures to prevent contamination from reaching the receptor or, in the case of a supply well, provide for an alternative water source.

Septa fitting. A special fitting used to seal vials (a liner for a threaded cap) or gas chromatographs (GCs) to provide closure. Septas can be manufactured in single, double, or triple layers of silicone rubber and other plastic materials. A syringe with a measured quantity of contaminant can be injected through a septa closure and into a GC column for separation analysis.

Sequential leach test. An extraction test method used to determine the leachability of metals from a solid matrix. Five successive chemical extractions are performed on the sample. Each extraction uses a leaching medium more aggressive than the previous medium (i.e., the pH of the leaching solution will vary from near neutral to highly acidic).

Sequester. To undergo sequestration.

Sequestration. Inhibition or stoppage of normal ion behavior by combining with added materials, especially the prevention of metallic ion precipitation from solution by formation of a coordination complex with a phosphate.

SESOIL. A one-dimensional model for estimating pollutant distribution in an unsaturated soil column. SESOIL results are commonly used to estimate the source term for groundwater transport modeling of the saturated zone.

Settling chamber. A basin or tank in which water or wastewater containing solids that can be settled is retained to remove by gravity a part of the suspended matter.

Shale. A fine-grained sedimentary rock formed by the consolidation of clay, silt, or mud. It is characterized by a finely laminated structure and is sufficiently indurated so that it will not disassociate when in contact with water.

Shale line (baseline). In electrical logging of a borehole using the spontaneous potential method, it is the line drawn through the extreme positive deflection on the SP curve.

Shear zone. A tabular zone of rock that has been crushed and brecciated by parallel fractures due to shear strain.

Shelby tube or split spoon sampler. Devices used in conjunction with a drilling rig to obtain an undisturbed core sample of the strata.

Short-circuiting. The entry of ambient air into an extraction well without first passing through the contaminated zone. Short-circuiting may occur through utility trenches, incoherent well or surface seals, or layers of high permeability geologic materials.

Shrinkage limit (SL). In soil mechanics, the boundary between the solid and semisolid states of a soil.

SI (International System of Units). A standardized system of units created by the International Organization for Standardization (ISU). This system of measurement is predicated on seven base units.

Quantity	SI Unit (Symbol)
Length	Meter (m)
Mass	Kilogram (kg)
Time	Second (s)
Electric current	Ampere (A)
Thermodynamic temperature	Kelvin (K)
Amount of substance	Mole (mol)
Luminous intensity	Candela (cd)

Sieve analysis. Determination of the particle-size distribution of a soil, sediment, or rock by measuring the percentage of the particles that will pass through standard sieves of various sizes.

Silica gel. A porous solid material consisting of silica manufactured in pellets of various sizes; it is made by treating sodium silicate (water glass) with sulfuric acid.

Silicate. Any of a broad range of mineral compounds comprised of one to six silica groups (SiO_2), arranged in either rings or chains.

Silicone. A linear polymeric structure derived from siloxanes by substitution of an organic group for the oxygen atoms above and below the silicone atom; chlorine halogen may also be included in the compound.

Silt. Particles of intermediate size between sand and clay.

Sinkers. Dense-phase organic liquids that coalesce in an immiscible layer at the bottom of the saturated zone.

Sinter. The heating of metal powders, clays, and similar earths to a temperature well below the melting point; this treatment results in softening.

Skewness. The lack of symmetry of an asymmetrical frequency distribution.

Slag. The non-gaseous waste material formed in a metallurgical or high temperature incineration furnace.

Slake. To become mixed with water so that a true chemical combination occurs.

Slickenside (soils). A smooth surface with parallel striae and grooves in the cutan between the soil peds. It occurs in soils containing appreciable quantities of expanding layer lattice silicates which are subjected to a monsoon type of climate.

Slug test. An aquifer test used to determine the *in situ* hydraulic conductivity of an aquifer by the instantaneous addition or removal of a known quantity (slug) of water into or from a well, and the subsequent measurement of the resulting well recovery time.

Slurry. A free-flowing suspension of fine solid material suspended in a liquid.

Slurry trenching. A subsurface cut-off or wall of low permeability placed near polluting waste source in order to capture or contain the resulting contamination.

Smectite. A commonly used name for the montmorillonite group of clay minerals. These clay minerals exhibit high shrink/swell properties and a high cation exchange capacity (CEC).

Snell's Law. A law describing the refraction of a wave at a boundary between two materials of different composition:

$$v_1/v_2 = \sin\alpha/\sin\beta$$

where

v_1 = velocity of layer 1.
v_2 = velocity of layer 2.
$\sin\alpha$ = angle of incidence in layer 1.
$\sin\beta$ = angel of refraction at layer 2.

Soap. Any product that imparts surface activity to water.

Soda ash. Common name for commercial sodium carbonate.

Sodium adsorption ratio (SAR). A relation between soluble sodium and soluble divalent cations which can be used to predict the exchangeable sodium percentage of soil equilibrated with a given solution. SAR is defined by:

$$SAR = \left[Na^+\right] / \frac{\int \left[Ca^{+2}\right] + \left[Mg^{+2}\right]}{2}$$

Sodium arsenite. A salt used in testing for residual chlorine; also known as sodium metaarsenite.

Sodium hydroxide. A powerful base, often called caustic soda, having the formula NaOH and readily dissociating in water solution to form Na^+ and OH^- ions.

Soil. The layer or mantle of mixed mineral and organic material penetrated by roots. It includes the surface soil (horizon A), the subsoil (horizon B), and the substratum (horizon C), which is the basal horizon and is limited in depth by root penetration.

Soil bulk density (dry). The ratio of the mass of dried soil to its total volume.

Soil bulk density (wet). The expression of the total mass of moist soil per unit volume.

Soil evaporation. The loss of water by evaporation into the atmosphere from water films adhering to moist soil grams.

Soil horizon. A layer of soil, approximately parallel to the surface of the soil, that exhibits characteristics produced by soil-forming processes.

Soil infiltration rate. Maximum rate at which a soil, in a given condition at a given time, can absorb water.

Soil micromorphology. The study of soil in the size range where optical enhancement is needed for the eye.

Soil moisture. Water of the soil zone which is divided by the soil scientist into available and unavailable moisture. Available moisture is water easily abstracted by root action and is limited by field capacity and the wilting coefficient. Unavailable moisture is water held so firmly by adhesion or other forces that it cannot usually be absorbed by plants rapidly enough to produce growth. It is commonly limited by the wilting coefficient.

Soil moisture content. The amount of water contained in the soil, generally expressed as a percentage. The percentage in moisture content is equal to the mass of water in the soil divided by the mass of dry soil and multiplied by 100.

Soil moisture potential (total potential). The amount of work per unit quantity of pure water required to transport reversibly and isothermally an infinitesimal quantity of water from a pool of pure water, at a specified elevation and at atmospheric pressure, to a pool of water identical in composition to the equilibrium soil solution (at the point under consideration), but in all other respects being identical to the reference pool.

Soil moisture tension. The equivalent negative pressure in soil water. It is equal to the equivalent pressure that must be applied to the soil water to bring it to hydraulic equilibrium through a porous permeable wall or membrane with a pool of water of the same composition. The pressures used and the corresponding percentages most commonly determined are 15-atmosphere percentage, 15-bar percentage, 1/3-atmosphere percentage, 1/3-bar percentage, and 60-centimeter percentage.

Soil physics. The science dealing with the physical properties of soil as well as the measurement, prediction, and control of the physical processes taking place in and through the soil.

Soil pore water. Water adhering to a soil grain in an unsaturated soil.

Soil pores. That part of the bulk volume of soil not occupied by soil particles; interstices; voids.

Soil salinity. The amount of soluble salts in a soil; the conventional measure of soil salinity is the electrical conductivity of a saturation extract.

Soil saturation. The filling of all the pore space in a soil with water.

Soil series. Soil horizons that are uniform in kind and arrangement.

Soil texture. (1) Relative proportions of the various soil separates in a soil material; (2) interrelationship between the size, shape, and arrangement of minerals in a soil.

Soil venting. An *in situ* remediation technique that enhances the migration of volatile constituents in the soil by artificially inducing differential pressures.

Soil washing/soil flushing. A physical treatment process which extracts contaminants from a sludge-soil matrix using a liquid medium process.

Soil water characteristic curve (soil moisture retention curve). An experimentally derived graph showing the soil moisture percentage (by weight or volume) vs. applied tension (or pressure). Points on the graph are usually obtained by increasing (or decreasing) the applied tension or pressure over a specified range.

Solid state. In physical chemistry, matter in the crystalline form.

Solonization (alkalization). The accumulation of sodium ion on the exchange sites of a soil.

Sols (suspensoids). Colloidal dispersions of solids in liquids.

Soluans. A coating of crystalline salts such as carbonates, chlorides, and sulfates on a soil.

Solubility in water (S_w). The water solubility of a compound is the saturated concentration of the compound in water at a given temperature and pressure. S_w is perhaps the most important factor in estimating a chemical's fate and transport in the aquatic environment. Compounds with high water solubilities tend to desorb from soils and sediments (i.e., they have low K_{oc} values), are less likely to volatilize from water, and are amenable to biodegradation.

Solute. A substance dissolved in a solution.

Solution. A uniform molecular or ionic mixture of one (or more) substance(s) (solute) in another (solvent).

Solution channel. A tubular or planar channel formed by solution in carbonate rock (Karst) terrains.

Solvent. A substance which dissolves other materials, reducing them to molecular or ionic form.

Solvent solubility. Concentration of a material that dissolves in a given solvent.

Sorb. A term used by chemists in cases when it is not clear whether adsorption or absorption is involved, or when both are occurring simultaneously.

Sorbent canisters. Gas-tight canisters typically filled with activated carbon (charcoal) for collection and transport of vapor samples. In the laboratory, the vapors are desorbed and analyzed to identify the organic compounds and quantify their concentration.

Sorbent tubes. Glass tubes filled with a sorbent material that reacts chemically with specific organic compounds. Based on the nature of the sorbent and the extent of the chemical reaction, organic compounds can be identified and their concentration quantified.

Sorption. A general term used to encompass the processes of adsorption, absorption, desorption, ion exchange, ion retardation, chemisorption, and dialysis.

Sparge. Injection of air below the water table to strip dissolved volatile organic compounds and/or oxygenate the groundwater to facilitate aerobic biodegradation of organic compounds.

Sparger. An air diffuser which is designed to produce large bubbles and is used singly or in combination with mechanical aeration devices.

Specific absorption. The capacity of water-bearing material to absorb water after all gravitational water has been removed. It is expressed as the ratio of the volume of water absorbed to the volume of material saturated.

Specific capacity. The rate of discharge of a water well per unit of drawdown, commonly expressed in gpm/ft or m^3/day/m. It varies with duration of discharge.

Specific conductance. The electrical conductivity of a water sample at 25°C (77°F), expressed in micro-ohms per centimeter.

Specific density. The specific density, also known as relative density, is

$$\text{Specific density} = d/d_w$$

where

    d = specific density of the substance.
    d_w = density of distilled water (g/mL or g/cm^3).

Specific discharge. For groundwater, the rate of discharge of groundwater per unit area of the porous medium measured at right angles to the direction of flow.

Specific gravity (G_s). The weight of a particular volume of a substance compared to the weight of an equal volume of water at a reference temperature, usually 4°C; also known as relative density.

Specific heat. The heat required to raise 1 gram of a material 1°C.

Specific permeability (intrinsic permeability). The property of a porous medium that measures the ease with which a fluid is transmitted through it under a hydraulic gradient.

Specific retention (S_r). The ratio of the volume of water that a soil or rock will yield by gravity to the volume of the material.

Specific storage (S_s). The volume of water released from or taken into storage per unit volume of the porous medium per unit change in head.

Specific yield (S_y). The percentage of the volume of water which a rock or soil, after being saturated, will yield by gravity to the volume of the rock or soil. It is expressed as:

$$S_y = V_w(\text{drained})/V$$

where

$$
\begin{aligned}
S_y &= \text{specific yield.} \\
V_w(\text{drained}) &= \text{volume of water drained from the rock or soil.} \\
V &= \text{total volume of rock or soil.}
\end{aligned}
$$

Average values of specific yield for unconsolidated sediments are listed below.

Unconsolidated Sediment	Specific Yield (%)
Clay	2
Sandy clay	7
Silt	8
Fine sand	21
Medium sand	26
Coarse sand	27
Fine gravel	25
Medium gravel	23
Coarse gravel	22

Specific volume. The volume of a unit mass.

Spectrophotometer. Instrument that measures the ratio of the radiant power of two electromagnetic beams as a function of spectral wavelength.

Spectroscopy. The branch of analytical chemistry devoted to determination of the structure of atoms and the chemical composition of molecules by measuring the radiant energy they absorb or emit in any of the wavelengths of the electromagnetic spectrum (gamma and x-radiation, ultraviolet, infrared, visible, microwave, and radio frequency).

Spectrum. A range of frequencies within which radiation has some specified characteristics, such as audiofrequency spectrum, ultraviolet spectrum, or radio spectrum.

Spent oxide. Bed material discarded from an oxide box when its activity for reacting with sulfur compounds has been lost.

Spline. A flexible strip of metal or plastic used by a draftsperson to draw a smooth curve between data points.

Spontaneous potential log (SP). Type of electrical logging in a borehole that provides a record of the naturally occurring electrical potential developed between the borehole fluid, formation water, and the surrounding rock materials as a function of depth.

Stabilization ponds. Large shallow basin used for purifying industrial wastes which encourages the growth of bacteria and algae in converting organic materials to nontoxic organic substances.

Stabilization/solidification. Remedial treatment that improves the structural stability and reduces the migration of waste; it often involves the conversion of liquid waste to a solid waste.

Stabilizer. Any substance added to another substance or mixture to prevent or retard a chemical or physicochemical change, at least for a considerable time.

Standard. A known reference chemical compound. The standard can be in a vapor state in a volatile organic analysis vial or in a Tedlar® bag. If a portable gas chromatograph is equipped with heated injection ports, standards in distilled water or methanol can be used. The concentration of the standard is usually known and, if so, can be used to withdraw a predetermined volume of the headspace gas. The headspace aliquot can then be injected onto the column for chromatographic analysis. Comparison of the retention times of the standard to the retention time of unknown sample peaks tentatively identifies the unknown sample peaks.

Standard atmosphere. A physical constant equal to 1.01×10^5 Newtons per square meter or 14.7 pounds per square inch.

Standard deviation. The positive square root of the variance, which is the average of the squares of the differences between the actual measurements and the mean.

Standard solution. A solution whose reacting value or strength per unit volume is known.

Star plot. A star plot is a statistical means to plot multivariate data so that different observations within a data set can be observed. A star consists of a series of rays drawn from the center, each ray representing one variable with the small value in each variable plotted as the shortest ray and the largest value the longest ray.

Static head. The height above a standard datum of the surface of a column of water (or other liquid) that can be supported by the static pressure at a given point.

Static pressure. The pressure exerted by a fluid in all directions at rest. It is the mean normal compressive stress on the surface of a small sphere around a given point.

Static water level. The water level in a well that represents an equilibrium condition when the aquifer is not being stressed (no nearby withdrawal or injection of water). Because groundwater conditions are generally dynamic, static is a condition that holds true only for short periods of time (anywhere from minutes to years), depending on cultural and climatic influences.

Statistical bias. The discrepancy between the expected value of an estimator and the population parameter being estimated.

Steady flow. Flow that is constant in magnitude, direction, and time.

Steam injection/vacuum extraction. A soil and/or groundwater remediation technique whereby injected steam is injected into the subsurface and recovers contaminants by increased volatilization, displacement of condensed liquids, and other mechanisms.

Steam stripping. A physical treatment process in which organic constituents are removed by volatilization through the heating of the matrix and injection of steam into the matrix.

STEL (short-term exposure limit). The maximum concentration for a continuous 15-minute exposure period to radiation.

Stephan-Boltzmann constant. A physical constant that is equal to 5.67×10^{-8} joule/$deg^4/m^2/s$.

Stephan-Boltzmann Law. A law that states that the total energy emitted by a body, J, integrated over all wavelengths is proportional to the fourth power of the absolute temperature, T. The law is often stated by:

$$J_t = \varepsilon\sigma T^4$$

where

J_t = total energy emitted by a body.
ε = emissivity coefficient.
σ = a constant.
T^4 = absolute temperature to the fourth power.

For a perfect "blackbody", the emissivity coefficient is equal to 1.

Sticky limit. The limit at which a soil loses its ability to adhere to a metal blade.

Stilling well. A pipe, chamber, or compartment with a comparatively small inlet or inlets communicating with a main body of water.

Stoichiometric. The exact proportions of elements in a chemical compound or reactant used to create a compound.

Stoichiometry. Study of the mathematics of the material and energy balances (equilibrium) of chemical reactions which is based on the laws of conservation of mass and energy and on the combining weights of elements.

Stoke's Law. A law used to describe the velocity of particles in a fluid. According to this law, the terminal velocity of a spherical particle settling under the influence of gravity in a fluid of a given density and viscosity is proportional to the square of the particle's radius.

Storage coefficient (S). Volume of water an aquifer releases from or takes into storage per unit surface area of the aquifer per unit change in head. In an unconfined aquifer, storage equals specific yield.

Storativity (S). *See* Storage coefficient.

STP (standard temperature pressure). A conventional abbreviation for standard temperature and pressure which is used by scientists for room-temperature research. The standard temperature is 25°C (77°F) and the pressure is 1 atmosphere (760 mm of mercury).

Strain. The measurement of the linear or volumetric deformation of a stressed material.

Stratification. A layered structure of sedimentary rock.

Stratigraphy. The science (study) of original succession and age of rock strata, also dealing with their form, distribution, lithologic composition, fossil content, and geophysical and geochemical properties. Stratigraphy also encompasses unconsolidated materials (i.e., soils).

Stratum. A single bed or layer of rock that is more or less homogeneous.

Stream (gaining). The term *gaining stream* replaces the term *effluent stream* and refers to a stream or reach of a stream whose flow is increased by the inflow of groundwater.

Stream (losing). The term *losing stream* replaces the term *influent stream*. It is a stream or reach of a stream that is losing water to the ground.

Streamline flow. A type of flow in which there is a continuous steady motion of the particles.

Stress. The force per unit area.

Stress cutan. Soil peds that press against each other during soil wetting.

Strike. The bearing of a horizontal line in an inclined plane.

Structural anomaly. A geologic feature, especially in the subsurface, distinguished by geophysical, geological, or geochemical means, which is different from the general surroundings.

Structural features. The group of geologic geometric features produced by deformation after deposition or crystallization; fractures, folds, and faults are examples of structural features.

Structure. In soil morphology, the aggregation of individual soil particles into larger units with planes of weakness between them.

Sublime. A term used by chemists to describe the change of a substance from the solid to the vapor phase without passing through the intermediate liquid phase.

Substituent. Any chemical entity that takes the place of another during the course of a replacement reaction.

Substitution (hydrolysis). A chemical reaction in which one substituent on a molecule is replaced by another.

Substrate. A term used by biochemists to denote any organic compound that is acted upon by an enzyme (e.g., a protein, carbohydrate, fat, or sugar).

Sulfonation. The addition of an SO_2OH group to an organic compound, the group being bonded directly to a carbon atom. The resulting compounds are called sulfonic acids (e.g., benzenesulfonic acid) and are formed by the reaction with sulfuric acid.

Sulfuric horizon. A subsurface soil layer which is mineral or organic and has a pH less than 3.5.

Sump. A pit or depression where liquids drain, collect, or are stored.

Supercritical water oxidation. Treatment process in which air is mixed with aqueous wastes above the critical temperature and pressure of water to oxidize organic wastes to carbon monoxide and water.

Supersaturation. A condition in which a solute is present in a solution beyond its limit of solubility.

Surface-active agent. Substances that reduce the interfacial or surface tension of liquids.

Surface area. Total area of exposed surface of a solid material, usually in the form of a powder, including all irregularities, and pores.

Surface drainage. Removal of water that accumulates on soil due to sloping land.

Surface impoundment. Facility or part of a facility which is a natural topographic depression, manmade excavation, or diked area formed primarily of earthen materials (although it may be lined with manmade materials). It is designed to hold an accumulation of liquid wastes or wastes containing free liquids and is not an injection well. Examples of surface impoundments are holding, storage, settling, and aeration pits, ponds, and lagoons.

Surface of separation. The surface separating displaced material from stable material but not known to have been a surface on which failure occurred.

Surface resistivity. A technique that measures relative values of the Earth's electrical resistivity. The technique is used to define subsurface geologic and hydrologic conditions.

Surface tension. The free energy in a liquid surface produced by the unbalanced inward pull exerted by underlying molecules upon the layer of surface molecules. This force contracts the surface to the minimum area.

Surfactant. A substance capable of reducing the surface tension of a liquid in which it is dissolved.

Surfactant injection. A form of subsurface remediation. One technique involves the injection of surfactant to create ultra-low interfacial tensions to reduce or eliminate capillary trapping forces and mobilize organic liquids within the pore spaces. This technique can also be used to enhance the solubility of a contaminant while minimizing interfacial tension reduction.

Suspended solids. Solids that are not in true solution and can be removed by filtration. They may be imparted from small particles of insoluble matter, from turbulent action of water on soil, or from domestic and industrial wastes.

Syringe blanks. Syringe blanks are used when air or headspace samples are being introduced into the instrument via manual injection. A known quantity of the ambient air is injected into the samples. Alternatively, syringe blanks may be obtained by withdrawing air from an empty pre-cleaned volatile organic analysis vial or other supply of clean air if the ambient air is suspected of contamination. Syringe blanks should be run at the beginning of each day and after every contaminated sample to ensure that no contaminants remain in the syringe before the next sample is run. Chromatograms for blanks should be void of any peaks.

Systematic sampling. A method in which measurements or samples are collected at locations and/or times according to a spatial or temporal pattern.

T

Tar. Residue obtained by destructive distillation of carbon-rich materials such as coal, wood, or petroleum.

Tare. The weight of a container used in gravimetric measurement.

Tedlar® bags. Gas-tight bags constructed of nonreactive material (Tedlar®) for the collection and transport of gas/vapor samples.

Tensiometer. A device used to measure the *in situ* soil water matric potential. A tensiometer consists of a porous, permeable material connected to a rigid tube that is attached to a manometer, vacuum gauge, pressure transducer, or other pressure measuring device.

Terminal electron acceptor (TEA). A compound or molecule that accepts an electron (is reduced) during metabolism (oxidation) of a carbon source. Under aerobic conditions, molecular oxygen is the terminal electron acceptor. Under anaerobic conditions, a variety of terminal electron acceptors may be used. In order of decreasing redox potential, these TEAs include nitrate, manganese, ferric iron, sulfate, and carbon dioxide. Microorganisms preferentially utilize electron acceptors that provide the maximum free energy during respiration. Of the common terminal electron acceptors listed above, oxygen has the highest redox potential and provides the most free energy during electron transfer.

Terne. A lead alloy having a composition of 10 to 20% tin and 80 to 90% lead which is used to coat iron or steel surfaces.

Tertiary. (1) A geologic period that includes the Pliocene, Miocene, Oligocene, Eocene, and Paleocene epochs; the Tertiary period began about 66.4 million years ago and ended 1.6 million years ago. (2) In reference to alcohols, this term denotes the presence of three alkyl groups attached to the methanol carbon atoms, and in reference to amines it denotes the attachment of three alkyl groups to the nitrogen atom.

Tetraethyllead. A toxic, organometallic liquid introduced as a gasoline anti-knock additive; its use makes possible octane ratings of up to 100 or more. It is made by the reaction of ethyl chloride with a mixture of sodium and lead.

Texture (soil). The interrelationship between the size, shape, and arrangement of minerals or particles in a rock.

Textural class. Refers to the overall textural designation of a soil.

Thaw. A weather condition that occurs when the temperature rises above the freezing point and ice and snow melt.

Theis method. A mathematical model used to analyze aquifer test data; the transmissivity and storativity of an aquifer are obtained with this method. It is an analytical solution for the radial flow of groundwater in a confined aquifer to a well. The equations for the determination of transmissivity and storativity are

$$T = 15.3 \ Q \ W(u)/s$$

$$S = T \ t \ u/360 \ r^2$$

where

T	=	aquifer transmissivity (L^2/T).
Q	=	pumping rate (M/T).
W(u)	=	well function.
s	=	drawdown in the well (L).
S	=	storage coefficient.
t	=	time.
u	=	pumping rate.
r	=	distance from the pumping well to the observation well (L).

Thermal conductivity. An intrinsic property of a substance that describes its ability to conduct heat as a consequence of molecular motion.

Thermal desorber. Describes the primary treatment unit that heats petroleum-contaminated materials and desorbs the organic materials into a purge gas or off-gas.

Thermal desorption system. Refers to a thermal desorber and associated systems for handling materials and treated soils and treating off-gases and residuals.

Thermal diffusivity. A change of temperature produced in a unit volume by the quantity of heat which flows through the volume in a unit time. This change occurs when a unit temperature gradient is imposed across two opposite sides of the volume.

Thermal efficiency. An expression of the effectiveness of temperature in determining the rate of plant growth.

Thinner. A term used by paint technologists to refer to low-viscosity liquids such as hydrocarbons naphtha, benzene, and turpentine which are added to oil-base paints to adjust their viscosity to a suitable point for application.

Thiol. Any of a class of organic compounds, either aliphatic or aromatic, which are structurally similar to alcohols but are characterized by the presence of an –SH (sulfhydrate) group instead of an –OH group.

Thixotropy. The property of some gels to repeatedly become liquid upon agitation and then gelling at rest. It is applied to certain types of solid-in-liquid dispersions such as clay-water pastes, sand saturated with water, concentrated paints, and colloidal gel structures.

Threshold dose. The minimum exposure dose of a chemical that will evoke a stated or nontoxicological response.

Threshold limit. A chemical concentration above which adverse health or environmental effects may occur.

Tide. The periodic rise and fall of the Earth's oceans and atmosphere.

Till. Unsorted and unstratified rock fragments, generally unconsolidated, deposited directly by and underneath a glacier without subsequent reworking by meltwater. Till consists of a heterogeneous mixture of clay, silt, sand, gravel, and boulders ranging widely in size and shape.

Titration. An analytical method for the quantitative determination of a substance in a solution. A measured amount of a reagent is added to the solution until a reaction, such as a color change, precipitation, or electrical measurement, occurs to indicate that the end point of the reaction has occurred.

TLV (threshold limit value). An estimate of the average safe airborne concentration of a substance. The TLV represents conditions under which it is believed that nearly all workers may be repeatedly exposed day after day without adverse effect.

TOC. Total organic carbon.

Toluene (C_7H_8). Aromatic compound whose molecule is comprised of a benzene ring in which one hydrogen is replaced by a methyl group. It is obtained either from coal-tar or from petroleum. It has many important applications in the chemical industries: it is a source material for benzene and phenol, a solvent for many resins used in paints, an explosives base (trinitrotoluene), and a source of polyurethane resins (disocyanates).

Tonguing. A diagnostic term used to describe a soil in which a white-colored intrusion at least 5 centimeters deep and 5 millimeters wide has penetrated into a clay or sodium-rich horizon.

Torque. The effectiveness of a force to produce rotation about a center, measured by the product of the force and the perpendicular from the line of action of the force to the center about which the rotation occurs. Torque is usually measured at 1-foot radii.

Torr. A unit of measurement of a state of vacuum. A torr is equal to 1/760th of a standard atmosphere or about 1 millimeter of mercury.

Tortuosity. Average ratio of the actual flow path in porous medium; it is the ratio of the average length of the flow path in the pores divided by the length of the sample.

Total alkalinity. The volume of acid required to react with the hydroxide, carbonate, and bicarbonate in a sample. It is a measure of the equivalent concentration of the cations associated with the alkalinity-producing anions in the solution, excluding hydrogen.

Total dissolved solids (TDS). A term expressing the quantity of dissolved material in a sample of water.

Total organic carbon (TOC). A measure of the carbon present in a sample as part organic compounds; it is one of the four RCRA indicators.

Total percent retained. In sieve analysis of soils, it is the weight in grams retained on a sieve divided by the total weight in grams of the oven-dried soil.

Total petroleum hydrocarbons (TPH). A measure of the concentration or mass of petroleum hydrocarbon constituents present in a given amount of air, soil, or water. The term "total" is a misnomer in that few, if any, of the procedures for quantifying hydrocarbons are capable of measuring all fractions of petroleum hydrocarbons present in the sample. Volatile hydrocarbons are usually lost in the process and not quantified. Additionally, some non-petroleum hydrocarbons may be included in the analysis.

Total recoverable petroleum hydrocarbons (TRPH). An EPA method (418.1) for measuring total petroleum hydrocarbons in samples of soil or water. Hydrocarbons are extracted from the sample using a chlorofluorocarbon solvent (typically Freon-113) and quantified by infrared spectrophotometry. The method specifies that the extract be passed through silica gel to remove the non-petroleum fraction of the hydrocarbons.

Total solids. The sum of dissolved and suspended solids.

Total stress. The weight of the overlying rock and water acting on an arbitrary plane through a saturated geological formation at depth.

TOX. Total organic halogens.

Toxic effect. Any change in an organism which results in impairment of functional capacity of the organism (as determined by anatomical, physiological, bio-chemical, or behavioral parameters); causes decrements in the organism's ability to maintain its normal function; or enhances the susceptibility of the organism to the deleterious effects of other environmental influences.

Toxicity. The inherent capability of a substance to cause adverse effects in human, animal, or plant life.

Toxicity characteristic leaching procedure (TCLP). An extraction test used to determine whether a waste meets the applicable technology-based treatment standards for land disposal.

Trace. A descriptive term used in geotechnical engineering which is equal to 1 to 10% by dry weight.

Trace elements. Elements occurring in minute quantities as natural constituents of living organisms. Trace elements include silver, lead, cobalt, iron, zinc, nickel, and manganese.

Tracer. A foreign substance mixed with or attached to a given substance for the determination of the location or distribution of the substance.

trans-. A chemical prefix that means "on the other side" or "beyond"; it is the opposite of *-cis*, which means "on this side". The prefix is used to indicate the position of substituent atoms or groups in relation to double-bonded carbons.

trans-**1,2-dichloroethene (*trans*-1,2-DCE)**. A liquid with a slightly acrid odor. Available data conflict on whether there is a significant difference in the toxicity from short-term exposure to *trans*-1,2-DCE vs. *cis*-1,2-DCE.

Transducer. A device that is actuated by power from one system and retransmits it, often in a different form, to a second system.

Transformation. In microbiology, the release of DNA by lysis of one bacterium and the uptake of this DNA by a second bacterium.

Transformation products. Chemicals released in the environment that are suscep-tible to several degradation pathways. These include chemical (i.e., hydrolysis, oxidation, dehydrochlorination, reduction, isomerization, and conjugation), photolysis, or photooxidation and biodegradation.

Transformer oil. A dielectric liquid used in transformers as a thermal and electric insulator. Various types include mineral oils, chlorinated hydrocarbons called askarels, and silicone oils.

Transmissivity (T). The rate at which water is transmitted through a unit width of an aquifer under a unit hydraulic gradient. Transmissivity values are given in gallons per minute through a vertical section of an aquifer 1 foot wide and extending the full saturated height of an aquifer under a hydraulic gradient of 1 in the English Engineering system. In the International System, transmissivity is given in cubic meters per day through a vertical section of an aquifer 1 meter wide and extending the full saturated height of an aquifer under a hydraulic gradient of 1. It equals the integration of hydraulic conductivity (K) across the saturated part of the aquifer perpendicular to the flow path multiplied by the thickness (b) of an aquifer.

Transpiration. The process by which water absorbed by plants, usually through the roots, is evaporated as water vapor into the atmosphere from the plant surface.

Transverse dispersion (D_t). The spreading of a solute in groundwater in a direction perpendicular to the bulk flow.

Travel time. The time it takes a contaminant to travel from the source to a particular point downgradient.

Tremie pipe. A pipe used to fill the annular space (the space between the soil and the outside of the well casing) from the bottom up when completing a well installation or when sealing an abandoned well.

Trend. The bearing of the vertical plane containing a line.

Trickling filters. A biological treatment system in which wastewater is trickled over a bed of stones covered with bacterial growth. The bacteria breaks down the organic waste into less toxic forms.

Trihalomethanes (THMs). All one-carbon compounds with three halogen atoms; most commonly, the THM compounds are chloroform ($CHCl_3$), bromodichloromethane ($CHBrCl_2$), and chlorodibromomethane ($CHBr_2Cl$). All four are priority pollutants.

Trip blank. A field blank which is transported to the sampling site, handled the same as other samples, and then returned to the laboratory for analysis in determining the QA/QC of sample-handling procedures. A sample bottle which is filled with pure water in a laboratory travels unopened to the field and back to the laboratory. It is usually employed to determine whether volatile organic compounds are inadvertently added to a sample in transit or in the laboratory.

Tripolyphosphates. Salts with $P_3O_{10}^{-5}$ anion. Most common is sodium tripolyphosphate ($Na_5P_3O_{10}$)

T-test. A statistical method used to determine the significance of difference or change between sets of initial background and subsequent parameter values.

Turbidimetry. A method to measure the difference between the transmission of white light through a finely divided suspension as compared to transmission through a standard suspension. It is used to measure the turbidity of a solution.

Turbidity. A measure of fine suspended matter in liquids. Suspended matter includes clay, silt, finely divided organic matter, plankton, and other microscopic organisms in water. The standard method for determination is the Jackson candle turbidimeter.

Turbine wheel. A rotor designed to convert fluid energy into rotational energy. Hydraulic turbines are used to extract energy from water as the water velocity increases due to a change in head or kinetic energy at the expense of the potential energy as the water flows from a higher elevation to a lower elevation. The fluid velocity tangential component contributes to the rotation of the rotor in a turbomachine.

Turbulent flow. The flow of a liquid past an object such that the velocity at any fixed point in the fluid varies irregularly.

Tyndall effect. A phenomenon named after the English physicist that studied it. It occurs when a beam of light passes through a colloidal suspension. Because colloidal particles have dimensions greater than the average wavelength of white light, the colloids interfere with the passage of this light. As a result, reflected light can be observed at right angles to the beam of light.

U

Udic moisture regime. A soil that is not dry for 90 cumulative days.

Ultisol. A soil order characterized by a region where the mean annual soil temperature is 8°C or warmer.

Ultrasonic. Frequencies above 20,000 cycles per second.

Ultraviolet. Wavelengths of the electromagnetic spectrum which are shorter than those of visible light and longer than X-rays (10^{-5} to 10^{-6} centimeter wavelength).

Unavailable moisture. Water held so firmly by adhesion or other forces that it cannot usually be absorbed by plants rapidly enough to produce growth. It is commonly limited by the wilting coefficient.

Unconfined aquifer. An aquifer where the water table is exposed to the atmosphere through openings in the overlying materials.

Unconfined groundwater. Water in an aquifer that has a water table.

Unconformity. In geology, a time break in a sequence of depositions or beds.

Uncontrolled sites. Sources of hazardous waste where the contamination is increasing or migrating; no removal procedures or remedial actions have been undertaken.

Underflow. Groundwater that flows beneath the bed or alluvial plain or a surface stream, especially in arid regions.

Underground storage tanks. Stationary devices, typically constructed of non-earthen materials designed to contain an accumulation of hazardous waste, typically petroleum-related products which are held underground.

Undisturbed sample (soil). A soil samples in which the material has been subjected to so little disturbance that it is suitable for all laboratory testing of the material's physical properties.

Uniform dispersion. A mixture of two or more substances (gas-in-gas, solid-in-solid, liquid-in-liquid, or solid-in-liquid) in which the proportion of components is exactly the same in all parts of the mixture.

Uniformity coefficient. A numerical expression of the variety in particle sizes in mixed natural soils, defined as the ratio of the sieve size on which 40% (by weight) of the material is retained to the sieve size on which 90% of the material is retained. This ratio was proposed as a quantitative expression of degree of assortment of water-bearing sand as an indicator of porosity. The value of the coefficient for complete assortment (one grain size) is unity; for fairly even-grained sand, it ranges between 2 and 3; for heterogeneous sand, the coefficient may be 30.

Unit weight (γ). In engineering geology, the weight of soil and water per unit volume; this term is used synonymously with bulk density.

United Soil Classification System. A soil classification system developed by Casagrande in 1948 with primary divisions based on grain size. A secondary division consists of groups and group symbols commonly used on borings logs. The group symbols and secondary divisions are as follows:

Group Symbol	Secondary Divisions
GW	Well-graded gravels and gravel-sand mixtures with little or no fines
GP	Poorly graded gravels or gravel-sand mixtures with little or no fines
GM	Silty gravels, gravel-sand-clay mixtures, plastic fines
GC	Clayey gravels, gravel-sand-clay mixtures with plastic fines
SW	Well-graded sands and gravelly sands with little or no fines
SP	Poorly graded sands or gravelly sands with little or no fines
SM	Silty sands and sand-silt mixtures with non-plastic fines
SC	Clayey sands and sand-clay mixtures with plastic fines
ML	Inorganic silts and very fine sands, rock flour, silty or clayey fine sands or clayey silts with slight plasticity
CL	Inorganic clays or low to medium plasticity, gravelly clays, sandy clays, silty clays, clean clays
OL	Organic silts and organic silty clays of low plasticity
MH	Inorganic silts, micaceous or diatomaceous fine sandy or silty soils, clastic silts
CH	Inorganic clays of high plasticity
OH	Organic clays of medium to high plasticity, organic silts
Pt	Peat and other highly organic soils

Unsaturated. A term used by chemists to indicate that an organic compound contains certain atoms (usually carbon), one or more of whose valences are not satisfied; this results in the formation of double or triple bonds within the molecule.

Unsaturated hydrocarbon. A hydrocarbon in which at least two carbon atoms are joined by more than one covalent bond while all of the remaining bonds are occupied by hydrogen; a hydrocarbon with one or more double or triple bonds between carbons. Can react with hydrogen (usually in the presence of a metal catalyst) to become more saturated.

Unsaturated zone (vadose zone). The area between the ground surface and the underground water table; interstitial spaces in this zone contain moisture (water) and air.

Upgradient. In the direction of increasing static head.

Upgradient well. One or more wells which are placed hydraulically upgradient (i.e., in the direction of increasing static head) and are capable of yielding groundwater samples that are chemically representative of regional conditions.

Upper explosive limit. The minimum concentration (vol% in air) of a flammable gas or vapor required for ignition or explosion to occur in the presence of an ignition source.

Uppermost aquifer. The geologic formation, group of formations, or part of a formation that contains the uppermost potentiometric surface capable of yielding a significant amount of groundwater to wells or springs and limited interconnection, based upon pumping tests, between the uppermost aquifer and lower aquifers. Consequently, the uppermost aquifer includes all interconnected water-bearing zones capable of significant yield that overlie the confining layer.

UV/ozonation. A chemical treatment process in which a liquid is simultaneously subjected to ozone and ultraviolet radiation. The ultraviolet radiation enhances the oxidation power of ozone and increases the reaction rate with the organic compound.

V

Vacuum draft tube. A narrow tube lowered into an extraction well through which a strong vacuum is pulled via a suction pump at the ground surface. Fluids (gas, water, and/or free product) are drawn into the draft tube and conveyed to the surface for treatment or disposal. Depending upon the configuration of the extraction system, the inlet of the draft tube may be either above or below the static level of the liquid in the well.

Vacuum filtration. A type of filtration process in which a mechanically supported, cylindrical rotating drum, covered by a filter medium, employs a center vacuum to draw water into the drum while the solids are scraped off of the filter.

Vadose zone (unsaturated zone). The zone containing water under pressure less than that of the atmosphere, including soil water, intermediate vadose water, and capillary water. This zone is limited above by the land surface and below by a saturated zone.

Valence. A whole number indicating for any element its ability to combine with another element.

van der Waals forces. Extremely weak forces of interaction between unexcited atoms or molecules of gases; they account for the fact that gases do not behave strictly in accordance with theory (i.e., their behavior varies slightly from that required by the ideal or perfect gas laws).

Vapor. Gaseous form of substances that are normally in the solid or liquid state (at room temperature and pressure).

Vapor density. The ratio of the mass of vapor per unit volume. An equation for estimating vapor density is derived from a varied form of the ideal gas law:

$$PV = MRK/FW$$

where

P = pressure (atm).
V = volume (L).
M = mass (g).
R = ideal gas constant (8.20575×10^{-2} atm · L/mol · K).
K = temperature (K).
FW = formula weight.

Vapor extraction. A soil remediation process in which a vacuum is applied to subsurface media to volatilize and remove volatile organic chemicals (VOCs). Hot air may also be used to enhance volatilization of VOCs.

Vapor pressure (vp). The pressure exerted by a vapor in equilibrium with its liquid or solid phase in a confined space. It provides a semi-qualitative rate at which a substance will volatilize from soil and/or water.

Vaporimeter. An instrument for measuring the volume or the tension of a vapor.

Vaporization. The process by which a substance such as water changes from the liquid or solid phase to the gaseous state.

Variance. The average of the squares of the differences between the actual measurements and the mean.

Variogram. A measure of the change in a variable with changes in distance.

Vector. An organism that transmits or acts as a carrier of parasites.

Venturi effect. The decrease in pressure in a pipe.

Venturi tube. A closed conduit or pipe used to measure the rate of fluids flow.

Vermiculite. A clay mineral in which potassium and magnesium are removed from the lattice structure during formation. The silica content is high while the aluminum concentration is low.

Vinyl chloride. A gas used to manufacture polyvinyl chloride and copolymers, in chemical synthesis, and in adhesives for plastics.

Viscosmeter. An instrument used to measure flow properties of a fluid.

Viscosity. The property of a substance to offer internal resistance to flow; specifically, the ratio of the shear stress to the rate of shear strain. This relationship is defined by Newton's law of viscosity, which states:

$$\tau = \mu \, dv/dy$$

where

τ = shear stress.
μ = viscosity.
dv/dy = velocity gradient.

Viscosity, absolute. The force that will move 1 square centimeter of plane surface with a speed of 1 centimeter per second relative to another parallel plane surface from which it is separated by a layer of the liquid 1 centimeter thick.

Viscosity, dynamic (coefficient of molecular viscosity, coefficient of viscosity). A coefficient defined as the ratio of the shearing stress to the shear of the motion of a fluid.

Viscosity, kinematic. A coefficient equal to the ratio of the dynamic viscosity of a fluid to the fluid's density.

Viscous flow. A type of fluid flow in which there is a continuous steady motion of the particles.

Visible radiation. Wavelengths of the electromagnetic spectrum between 10^{-4} and 10^{-5} cm.

Vitreous. A term used in ceramic technology to describe matter in the glassy state.

Vitrification. An *in situ* thermal treatment technology which converts soil into a chemically inert and stable glass and crystalline matrix. Organic constituents within the soil are destroyed by pyrolysis.

Void ratio. A ratio that expresses the volume of soil pores to the volume.

Voids. The open spaces between solid material in a porous matrix.

Volatile. Capable of being evaporated at relatively low temperatures.

Volatile acids. Fatty acids containing six or less carbon atoms, which are soluble in water and which can be steam-distilled at atmospheric pressure.

Volatile constituents. Solid or liquid compounds which are relatively unstable at standard temperature and pressure and undergo spontaneous phase change to a gaseous state.

Volatile organic compounds (VOCs). Hydrocarbon-based chemicals that are characterized by low boiling points and high vapor pressures.

Volatile organics. Liquid or solid organic compounds with a tendency to pass into the vapor state.

Volatilization. The evaporation or changing of a substance from a liquid to vapor. If the chemical is an organic compound, it is called a volatile organic compound (VOC).

Volume wetness (θ). A ratio of the total volume of water to the total soil volume; synonymous with volumetric water content of volume fraction of soil water.

Volumetric water content (θ). A percentage of the total volume of water to the total volume of soil.

Vughs. Unconnected voids with irregular shape and walls.

W

Wafer. A thin disk or slice of silicon on which many separate chips can be fabricated and then cut into individual die.

Wash. Alluvium that has been collected, carried, and deposited by the action of water.

Wastage. The processes by which glaciers lose substance; these processes include wind erosion, corrosion, calving, evaporation, and melting and ablation.

Waste heat. A generic term used to describe heat that is rejected into the environment.

Waste minimization. The reduction of hazardous waste to the fullest extent feasible.

Water. An inorganic substance whose formula may be written either H_2O or HOH; it exists in three forms or phases: (1) as a crystalline solid (ice) at or below $0°C$ ($32°F$); (2) as a colorless liquid from 0 to $100°C$ (32 to $212°F$); and (3) as a vapor (steam) at $100°C$ ($212°F$).

Water content (w). The amount of water stored within a porous matrix as expressed on a volume per unit volume or mass per unit mass of solid basis. It is expressed as:

$$w = W_w/W_s \text{ or } M_w/M_s$$

where

w = water content (%).
W_w = weight of water.
W_s = mass of solids.
M_w = mass of water.
M_s = mass of solids.

Water gas. A gas rich in hydrogen, carbon monoxide, and methane (heating value ranging from 500 to 1000 Btu per cubic foot) made by pyrolyzing oil in the hot gas product from a blue gas generator. Also known as carburetted blue gas or carburetted water gas.

Water level recorder. A device for producing, graphically or otherwise, a record of the rise and fall of a water surface with respect to time.

Water loss. The difference between the average precipitation over a drainage basin and the water yield from the basin for a given period.

Water spreading. The artificial application of water to lands for the purpose of storing it in the ground for subsequent withdrawal.

Water solubility. The maximum concentration of a chemical that dissolves in pure water at a specific temperature and pH.

Water table. The water surface in an unconfined water body at which the fluid pressure in the pore of the porous media is at atmospheric pressure. It is defined by the levels at which water stands in wells that penetrate the water body just far enough to hold standing water. In wells which penetrate to greater depths, the water level will stand above or below the water table if an upward or downward component of groundwater flow exists.

Water table contour. A line connecting all points on a water table; also called phreatic decline.

Water table map. A contour map of the upper surface of the saturated zone.

Water yield. Runoff from the drainage basin, including groundwater outflow, that appears in a stream plus groundwater outflow that bypasses the gauging station and leaves the basin underground.

Watercourse. A natural channel through which water flows.

Watershed. A region drained by a stream or body of water.

Wax. A thermoplastic solid of high molecular weight which may be a hydrocarbon derived from distillation of an aliphatic petroleum (paraffin wax) or an ester of unsaturated fatty acid and alcohols which are end products of plant and animal metabolism.

Weathering. Mechanical, chemical, or biological disintegration and decomposition of rocks.

Weighted average. A statistical means by which the degree of importance to a value is assigned. A weighted average is calculated by multiplying each data value by the appropriate weighing factor and dividing the total by the sum of the weighing factor.

Well head. The area immediately surrounding the top of a well, or the top of the well casing.

Well log. A record of installation of a well. It includes construction specifications of the well, depth, owner, location, and a description of the soil profile and is prepared by the well driller.

Well point. A screening device, equipped with a point on one end, that is driven into the ground.

Well purging. The process of removing water from a well to allow *in situ* formation water to enter the well.

Well screen. That portion of the well casing material that is perforated in some manner so as to provide a hydraulic connection to the aquifer.

Well surging. That process of moving water in and out of a well screen to remove fine sand, silt, and clay-size particles from the adjacent formation.

Well yield. The volume of water discharged from a well in gallons per minute or cubic meters per day.

Wet air oxidation. A thermal treatment process that breaks down organic materials by oxidation in a high-temperature and -pressure aqueous environment in the presence of compressed air.

Wet oxidation. A thermal treatment process in which organic materials are degraded through the use of elevated temperatures and pressures in a water solution or suspension.

Wetting agent. An additive which reduces surface tension.

Wetting front. Movement of a fluid through the soil to the point that the moisture gradient is so steep that there appears to be a sharp boundary between the moisture content above and below this boundary.

Wien's Law (λ_m). A law that states that the maximal radiation intensity is inversely proportional to the absolute temperature. Wien's law is expressed as

$$\lambda_m T = 2900$$

where

λ_m = the wavelength, in microns.
T = temperature, in Kelvin.

Wilting coefficient. The ratio of the weight of water in the soil when the leaves of plants first undergo permanent reduction in their water content as the result of a deficiency in the supply of soil moisture to the weight of the soil when dry.

Windrow. A low, elongated row of material left uncovered to dry; windrows are typically arranged in parallel.

Wipe sample. A sample collected for the purpose of determining the presence of removable contaminants from a surface; sampling is usually performed by wiping the surface area with slight pressure with a piece of soft filter paper or cloth.

White spirits (petroleum spirits, mineral spirits, or Stoddard solvents). A hydrocarbon derived from the light distillate fractions during the crude-oil refining process. They are composed of the C_6 to C_{11} compounds, with the majority of the relative mass composed of C_9 to C_{11}. White spirits are composed of the following general classes of compounds: 50% paraffins, 40% cycloparaffins, and 10% aromatics.

Withdrawal. Water pumped out of a well.

X

Xerophytes. Plants that grow in or on extremely dry soils or soil materials.

X-ray absorption. A method of analysis used to measure the presence of heavy elements in a substance that is composed primarily of low-atomic-weight materials.

X-ray diffraction. A method of analysis commonly used to study minerals in which X-rays are reflected off the surface of a crystal. As the crystalline material is rotated within the path of the radiation, information concerning the structure of the crystal can be obtained.

X-ray emission (X-ray fluorescence). A method of analysis in which a sample is bombarded with high-energy X-rays which are absorbed by certain elements and re-emitted as lower energy X-rays. The lower energy radiation emitted is unique for different elements.

Xylene. A derivative of benzene in which two of the hydrogen atoms are replaced by methyl groups: $C_6H_4(CH_3)_2$.

Z

Zeolite. A group of hydrated aluminum complex silicates, either natural or synthetic, with cation-exchange properties.

Zone of saturation. The zone below the water table in which all interstices are occupied by groundwater.